Envelope Design for Buildings

This book is dedicated to the memory of
Robert Fitzmaurice,
great and far-seeing communicator, friend and mentor

Envelope Design for Buildings

William Allen

Architectural Press

Architectural Press
An imprint of Butterworth-Heinemann
Linacre House, Jordan Hill, Oxford OX2 8DP
A division of Reed Educational and Professional Publishing Ltd

A member of the Reed Elsevier plc group

OXFORD   BOSTON   JOHANNESBURG
MELBOURNE   NEW DELHI   SINGAPORE

First published 1997

British Library Cataloguing in Publication Data
A catalogue record for this book is available from the British Library

ISBN 0 7506 2854 5

Library of Congress Cataloguing in Publication Data
A catalogue record for this book is available from the Library of Congress

Composition by Scribe Design, Gillingham, Kent
Printed and bound in Great Britain

Contents

Foreword

Architecture is the most important of the visual arts. Inside or out you cannot avoid it. Its success relies on the creative and imaginative use of space and materials. Nevertheless, however successful the architecture, if the roof leaks, the wind whistles through ill-fitting windows, and the interior environment is uncomfortable it has failed its primary requirement of providing shelter from the elements. In the past when building was relatively simple, relying on a limited number of well-tried and tested materials in the hands of skilled craftsmen, serious errors in design specification or construction were rare.

Today buildings are bigger and far more complex than ever before. Design now requires a very high degree of expertise and understanding of a far greater range of materials and how they are put together to ensure they function as required. The possibilities of things going badly wrong is greater now than ever before.

This book, written by one of the most experienced architects in the performance of buildings is not, however, about defects. It deals comprehensively with designing the building envelope, the element which separates the external climate, over which we have little control, and creating the right interior environment. Covering roofs, walls, basements; new materials and old; and how they fit together and react to each other, all the factors involved with design and specification which affect the construction and ultimate performance of the building are examined here. This is a book for all concerned with the design of buildings: clients, architects, constructors and students. Follow the principles and guidance in this book and the quality of the architecture should not be denegrated by the failure of the building to provide comfort and shelter against the elements.

Owen Luder CBE
President RIBA

A personal note and preface

I came from Canada to Britain just before the war seeking involvement in the design of concert halls, theatres and the like, for I had a vision of the way design had been and should be used to enhance the beauty of sound. But first I had to have a better working knowledge of acoustics myself and on the advice of the late Hope Bagenal, architect, friend, and the greatest acoustic consultant of his time, I accepted an invitation to joint my dedicatee, Robert Fitzmaurice, at what was then the Building Research Station, for among his responsibilities was acoustic research.

The Station – now Establishment – was then less than 20 years old and had pioneered building science. Fitzmaurice himself had just seen publication of Volume 1 of his classic work, *The Principles of Modern Building*. It caught at once the imagination of those leading young moderns of the day, F. R. S. Yorke, Frederick Gibberd and Wells Coates, who saw in it the foundation elements for their architectural beliefs. Donald Gibson, later to replan Coventry and to develop CLASP for school construction, was, like me, on Fitzmaurice's staff. Fitz, as he was universally known, was a charismatic figure, and all these as well as bright young builders and engineers circled round him constantly like electrons around a nucleus.

Volume 2 of *Principles* was already on the stocks and I was invited to broaden my interests by helping Fitzmaurice to develop it. This was to include roofs and he explained that because he liked to have historical roots for discussions of modern technology, he had made a list of about twenty pubs up and down the country which he knew to have particularly good ancient roofs and he proposed that we should approach our task by a study of these. This admirable way of going about research has been a life-long influence upon me.

Volume 2 was overtaken by the onset of war and when five years later reconstruction had to be faced, it was soon evident that it was only the bright-eyed, bushy-tailed young moderns of the late 1930s who had the necessary attitude, aptitudes and inner beliefs to handle it. Gibson, setting a lead with Coventry, and Percy Johnson-Marshall, fresh from the fray in the Far East, inspired the entire country with determination to create a better built environment for the future, while Percy's brother Stirrat, also fresh from the East and from Normandy, took on the key role for the provision of better (and cheaper) schools, initially and famously in Hertfordshire and later from the central seat in the Ministry of Education. It was a highly creative moment and full of lessons for any country, now or in the future, which has to replace extensive destruction in the aftermath of war or natural disaster.

The pre-history of this book

Before Fitzmaurice began to write *Principles* he had realized that the prescriptive form into which building regulations had been cast from time immemorial belonged to the pre-scientific era, for they controlled so rigidly the way buildings were built that they precluded the acceptance into practice of advances that could be expected to flow from building research. He therefore proposed that they should be reformulated on a performance basis to open the way to justifiable innovation.

The government agreed in principle and it was an historically important break with tradition by which construction and, in some respects, architecture itself could be opened up to full participation in the age of science.

Principles was intended to put exemplary findings from early building research into the hands of design professionals and the industrial littoral of building, but forward movement ceased during the hostilities and when they were over, no time was found for administrative change. House replacement and schools were the priorities and their numbers had to be expanded much beyond the capacity of the recovering conventional industry, so both had to be turned towards innovation, for which performance criteria were suddenly required from building research. The position which Fitzmaurice had expected to take several years to reach occurred almost overnight and it was more than a decade before the formalities of the new control methodology caught up with it.

In the early 1950s the priorities for houses and schools were relaxed and a more normal spectrum of building types began to appear, including quite a large proportion in high-rise construction for low-cost dwellings. Innovation was the order of the day and soon went well beyond the knowledge base needed for it, though this was not apparent until the late 1960s, when failures began to make themselves evident.

Historically it was a moment when intensive research-linked monitoring should have been organized so that learning and feed-back could take place quickly and systematically, but historical awareness was nowhere evident and instead learning and feedback took place around individual misfortunes which were investigated as isolated incidents by an unorganized mix of building researchers and interested architectural and engineering practices. Co-ordination and feedback was mostly limited to seminars and was neither quick or efficient. Educators, unfortunately, were not in much evidence.

The incidence of failures became very high, running eventually to quite a few thousands, and they became very varied. There was also a good deal of litigation as aggrieved parties and insurers struggled for redress. It was an unavoidable activity and had one outstanding merit, for it required the depth of investigation necessary for proof and had therefore a high learning content.

The architectural firm which the late John Bickerdike and I founded together in 1962 was soon drawn into this work, most probably because it was known that both of us had been at the Building Research Station. We felt able to to do the work, we felt a sense of duty about it, and we saw educational advantages for the practice, so we allowed it to develop alongside our more normal activity. Because we were frequently asked to carry out the remedials, we had to think positively about the use of what we were learning as feedback. This became one of the roots of the present book.

It must also be said that while most of the misfortunes happened to post-1945 buildings, historic structures did not escape. Changes of materials, changed requirements – notably of thermal insulation – and changes in habits of construction that seemed innocuous but were not, have taken a toll and also made their contribution to the learning process. Misplaced thermal insulation has been particularly destructive. The cost of all this cannot be established but it runs to many hundreds of millions of pounds and almost certainly to over a billion, accompanied by much human misery.

But this is not a book about misfortunes. It is what is learnt from them that matters, and what I have attempted to do here, as a result, is to think through the technology of building envelopes afresh as a process of design decision making. It seemed wrong to have had such privileges of education without doing so.

Preface to the book itself

The book is about concepts, processes and ideas, and looks at the way things work, because to my designer's mind that is what I have to understand for problem solving in design, for innovation, and for reliability. It is not a book of finite design solutions – that would be alien to modern needs – and numbers mainly appear only when needed for explanation. They can easily give false confidence and mislead you if you do not maintain expert familiarity with their use and limitations. Judgement is an inescapable duty for professionals.

The book is divided into two parts, of which Part 1 deals with the external and indoor climates, while Part 2 examines the envelope systems which mediate between them,

In respect of the weather, our Anglo-Irish archipelago is said to enjoy a temperate climate. It is true for people but not for buildings, for which it is almost uniquely intemperate in the world. In a new look at climatic factors I have therefore pulled out those which have a bearing on design decisions and explained why they matter. Hopefully it is also an analysis which will help designers working in other climates to evaluate their influence more perceptively.

Indoor climates on the other hand are mainly begotten by what goes on indoors and by the characteristics of the enclosing envelope. In the period which generated this book there was one devastating climate change indoors in Britain which created an intensive learning demand, a country-wide epidemic of severe condensation in post-war housing. It was daily or weekly headline news from 1967 for three or four years and it struck in a few other occupancy types as well. It still occurs sometimes now. It was even a major factor in decisions to demolish some high-rise blocks of flats, though some demolitions were unnecessary if cause and cure had been better understood. I tell the very instructive story of why it all happened, and what to do about it design-wise, and I explain the difference needed in envelopes according to whether the indoor climate is governed by natural forces or by powered air changing.

In Part 2 one might say with Mies, God is in the details, and I look first at basements where so often in the past what happens below ground has been assumed simply to be an act of God and not subject to rational prediction. That time is definitely behind us and a more normal knowledge base exists for design, forming the substance of Chapter 3.

Moving above ground, the first wall systems discussed are a group based upon the use of a cavity. Originally cavities came into being when conventional wall brickwork had been thinned to nine inches and was then discovered to be leaky; the nine inches were split in two, separated by a cavity to prevent rain getting to the inner half.

After the war energy-consciousness developed, later accentuated by the world oil crisis, and increases of thermal insulation began to be required. The cavity provided a convenient location for it but altered at once the thermal behaviour of the two enclosing leaves, creating troubles in the outer while offering benefits to the inner. Then, on occasions when the outer brickwork was replaced by concrete cladding, its lower permeability to air and water changed conditions in the cavity in one direction while rainscreen systems, with total permeability to both, changed them in another. In the relevant chapter these and other variants are described along with their implications for design.

Various finishes such as tiling, mosaic, thin stone, and others came into use as designers explored options that had newly opened up for them, but little, if any, useful guidance was provided for their successful introduction and numerous misfortunes befell them, sometimes reflecting undeservedly upon their merits as products. Occasionally they were imports from less exacting climates. They need, and have therefore been given, a chapter to themselves.

Curtain walls follow the external finishes. As an architectural idea they have been lurking in the shadows of conceptual evolution for five centuries but had to await the technology of the nineteenth century to begin the spurt which metal framing made possible, and that of the twentieth for the maturity which now seems to be approaching. But in all curtain walling the technology of success is exacting, more so than that of any other enclosure system, and has often proved elusive. Never underestimate the need to understand it.

Timber-framed walling for low-rise property is not widely used in our archipelago and there have been a few problems. Such as they were, they have been dealt with in a brief chapter.

Glass and glazing, doors and windows, with hundreds of years of successful use behind them cannot have seemed likely to be accident prone, but they were. Sometimes the glass itself came to grief; sometimes it was used in ways it could not accept. Double glazing, inherently a bundle of thermal conflicts, forseeably ran into difficulties. But it was door and window frames, mostly those of timber, that caused the greater surprises. Some just fell apart, many leaked, and there was quite a high incidence of rot, induced by the paint industry's shift to a near-impermeable formulation that prevented damp timber from drying. Some of the problems have been overcome but there are many lessons that need to be kept on the record.

Flat and pitched roofs come last. Flat roofs failed on a majestic scale, largely because the rising requirements for thermal insulation had to be met by insulants that had to be kept dry. They had therefore to be placed under the weather membranes, an arrangement which caused the membranes to suffer hugely increased temperature ranges for which the conventional felts and asphalts had never been engineered, and they suffered much distress. Plastics entered the market, but not entirely without troubles of their own. The real solution to the difficulties came with waterproof insulants that can be placed over membranes, for these then enjoy greatly enhanced longevity from the protection they thus get.

A very great gain which has not yet been sufficiently recognized or exploited has resulted from the new-found ability to wrap the inner fabric of buildings in thermal insulation, using its protected positions in wall cavities and under the weather membranes of flat roofs, for it then enhances valuably the thermal stability of interiors, reducing or removing the need for structural breaks, and minimizing the thermal staining that leads to more frequent needs for redecoration. Among its potential benefits is the energy conservation made possible by thermal stability and the reduced heating and cooling plant capacity needed.

Penultimately I discuss pitched roofs in Chapter 10. The great longevity of tiling systems and metal roofing in Britain and Ireland was suddenly undermined by the unguided introduction of thermal insulation. Set up closely to the underside of tiled roofing it prevented the free circulation of air needed to keep the timber dry and, as usual, it enlarged the temperature range to which the tiles themselves were subject, and much the same mechanisms operated under metal roofs where the reasons for their longevity seem not to have been understood. Lead, in particular, suffered and is still suffering. It has often seemed natural now to place it on plywood, with insulation immediately beneath, but the conditions for rapid corrosion were then in place. Condensate formed under the lead and wetted the plywood, creating a catalytic situation chemically resembling the ancient methods of making white lead for paint. One such case is quoted where lead of a weight suitable for cathedrals failed in less than a year.

Such, therefore, is the structure of my text, but there is a note I should add about my handling of linear dimensions, for I sometimes use centimetres. The building industry in Britain chose to exclude them in its use of metrication, retaining only millimetres and metres. I think this sometimes suggests a level of accuracy not realistic on building sites and I have therefore used centimetres for approximations, as I would have used inches, when I thought it helpful to judgemental needs. Also, they are standard practice in countries where metric measurement has long been the norm and where I am told that readers of this book would not find their omission helpful.

And finally a general comment about materials. Almost universally they are made now by firms with development laboratories competitively seeking marketable improvements, cost reductions or new products, so the materials industry is not in a stable state. Change is inherent in science-based development and mostly it results in gains, but laboratories are not staffed by people from the real world of building, so sometimes lapses will occur and there will be moments of pain. I hope my offering helps both the design side and the industry to sort them out quickly.

The main text of the book is followed by two appendices, one on the thermo-moisture movement of building products and the other on background information on polymers and plastics. This latter piece has kindly been provided by Mr Arthur Burgess, for several years an invaluable consultant on chemical and other scientific matters to Bickerdike Allen Partners.

Acknowledgements

Over the long period it has taken me to get this book written I have become indebted to many people. My then secretary, Janet Wiesner, patiently typed the developed text and most of its predecessors and kept other demands on my time under strict control. Other secretaries also helped and have my thanks. Alaine Hamilton, also of BAP and later of RIBA Publications, herself an experienced author, gave me help of many kinds. My external referees, Professors Peter Carolin of Cambridge and Derek Poole of Cardiff were admirable critics and urged me on, and I had valuable specific help from Leo Biek, Dr William Bordass, and Ian Reith. The partners and associates of Bickerdike Allen Partners, including Peter Rich who retired early, and John Bickerdike, who died much too early, have all contributed to my education and have my deepest gratitude. Legal friends helped often to shape points of view. A 30-year consultancy to the American Building and Fire Research Division of the National Institute for Science and Technology kept me valuably in touch with American practice.

And finally I thank my wife, Tessa, who suffered at second hand and in many ways from the agonies and strains which have afflicted me in my struggles to put this book together. She ruefully laughed her way through the lot.

In the end, of course, what I have offered here are my own interpretations of what has come my way in the experience of this intensely educative period for architecture and the building industry, and I hope it will contribute usefully to both fields and to conservation.

William Allen
May 1997

Part One
Outdoor and indoor climates in Britain

1   The outdoor climate

1.1   Introduction

England, Ireland, Scotland and Wales occupy an archipelago of islands over which a distinctive climate prevails, and because it generates similarly distinctive influences on building design and behaviour, the body of experience upon which this book is founded relates parts of the text distinctively to these islands and to other parts of the world having climates of at least some similarity. Technically, the climate is classified as Temperate Oceanic, but despite the moderation which 'temperate' implies, it is as severely hostile to buildings as any climate in the world.

Let us begin with temperature. Every year, freezing occurs to some extent everywhere in England, Wales and Scotland, and in many parts of Ireland. Air temperatures sometimes drop as low as −10°C near the sea and lower in most inland areas. Every few years, freezing occurs for prolonged periods except in areas exposed to the Atlantic current from the south-west. Freeze/thaw cycling is therefore endemic, and because the coldest weather usually follows wet, freezing contributes significantly to degradation on building envelopes.

At the other end of the spectrum, air temperatures rise to 25°C or more in most areas on at least a few days every year, and occasionally for long periods. In prolonged hot spells the daytime air, together with sunshine, causes severe drying of the ground and the outer parts of the building fabric. The most extreme surface temperatures usually occur when clear skies coincide with cold, still nights or hot, windless days, the low due to radiation loss to the sky and the high due to radiation gain from sunshine. The greatest extremes occur on materials of low thermal capacity with insulation behind or beneath them and finished in a dark colour; the seasonal range on these often exceeds 100°C.

Gales, which are defined specifically as winds of 17 m/s (metres per second, about 35–40 mph) persisting for at least 10 minutes, occur on average two days a year in southerly and inland areas, and more than 30 days in the north and coastal areas. Gusts of at least 40 m/s (about 90 mph) occur at least once in 50 years everywhere, and in exposed positions may reach 55 m/s (120 mph) or more. Wind speeds are generally highest in the wettest areas and at the wettest times of year because they are a by-product of weather producing these extremes.

These and other weather data mostly relate to altitudes near sea level. Typically air temperatures drop about 0.6°C for every 100 m increase in altitude, while wind and wind-driven rain become more severe.

There will be few references in the book to 'average' weather. What we will focus on are the weather conditions which can adversely affect the fabric of the building envelope, and these are the severe combinations of temperature, sunshine, wind and water.

We cannot profile other climates here; they are too numerous and varied, and comparably relevant data for design often are not available. But the design professions and the construction industries now practise globally and hopefully will find the present profile useful at least in identifying what to look for in other people's weather and how to think about it in relation to design and construction.

1.2   The changeability of the weather

Local weather is the product of constantly moving weather systems, so let us first understand what weather systems are. In effect they are large air masses within which temperature differences develop, and these in turn create pressure differentials that seek equilibrium by rotational movements of high pressure to low. Sometimes the temperature differences are such as to form precipitation, and the rotational movements are what we experience as wind. Stability always eludes weather systems.

Our particular climate is largely the result of the location of these islands in a belt of globe-encircling air which is subject to exceptional disturbance. The eastward rotation of the earth drags all air and weather systems with it by friction, but along our latitudes there lie the huge plains of Siberia, the water mass of the North Pacific, the great plains of northern North America and another water mass, the North Atlantic. All have different temperatures at any time of year and all create big thermal surges in the air above them, while between the Pacific water and the North American plains lie the high Rocky Mountains, big enough to make the weather turbulent. Then, north and south of our weather belt are quasi-static bodies of polar and tropical air, which more or less channel the mix but also get caught up in it from time to time along its fringes, sometimes very influentially. The weather systems follow the earth's rotation relatively slowly, taking perhaps two weeks to complete an encirclement at our latitude.

The pressure differentials are displayed as the highs and lows on weather maps, surrounded by patterns of the intervening isobars (Figure 1.1). The rotary air movements take place along the gradients from high to low, forming large vortices at the lows which act rather like water running out of a bath and which have become a familiar feature of TV satellite weather pictures. The size of rotations varies a great deal: some are small and regional while others may spread over an area the size of western Europe. Hurricanes and cyclones are extreme forms of such rotations, but in our latitudes the conditions for their generation are not often present, although we sometimes get the tail-ends of hurricanes which afflict the eastern United States.

The closer the isobars are to one another, the steeper the pressure gradient will be, and the stronger the winds. The rotations are initiated by waves in the upper atmosphere created by the spin of the earth, the winds typically rotating clockwise from the highs and then spinning anti-clockwise into the vortices of the lows. The pressure differentials between the two lie typically in the range from about 1050 millibars (mb)

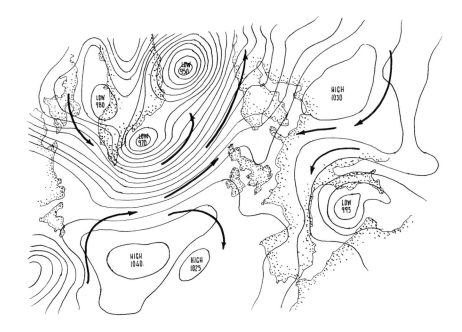

Figure 1.1
An example of a north Atlantic weather chart showing wind rotations. (Reproduced from material supplied by The Met. Office).

in the middle of a high to around 950 mb in the eye of a low. The greater the difference and the steeper the gradient, the stronger are the resulting air movements, i.e. the winds. A 'low' as low as 950 mb usually creates some that are very strong. In 1993, depressions as low as 920 mb occurred in the north of Scotland and resulted in winds of very high velocity. In a normal year 160 to 180 lows move across Britain, mostly on northerly tracks.

Wind direction over the surface of the land is thus determined by the tracks of the highs and lows, and because many lows track across the northern parts of the islands or between them and Iceland, all our land masses are often swept by north-westerly to south-westerly winds. Some lows wander across western and southern Europe, however, and bring us southerly, easterly or northerly winds. The lines of the tracks are influenced by what are termed jet streams in the upper atmosphere.

The air movements in the rotations do not all have the same temperature because of the general turbulence and because warm tropical and cold polar air masses get stirred in along the way. Sometimes warm air gets overtopped by cold air and sometimes the opposite happens. Either way, the layer of air at the boundary between these forms what is termed a 'front', either warm or cold depending largely on which air mass is dominant. Cold fronts are typically associated with precipitation.

For reasons that are not clear, the generally eastward movement of these weather systems is sometimes arrested, probably by conditions in the upper atmosphere. When this happens, spells of one kind of weather, wet or dry, may prevail for days, weeks or even months. The three or four dry summers that culminated in the drought season of 1976 are a well-remembered example. Such droughts seem often to end in heavy rain.

In all these ways, all parts of these islands become subject to cold or hot and wet or dry periods that last long enough to produce very damaging stress changes in the outer parts of building envelopes.

1.3 Winds

Most of the air that moves around over the British Isles has also recently passed over the large water mass of the Atlantic, which explains much of the prevalently high humidity and heavy rainfall in Ireland and on the western sides of England, Wales and Scotland. In addition, these mountainous areas induce precipitation by thrusting incident air upwards into a chillier atmosphere where vapour in it condenses, so that eastern England and Scotland and some areas of Ireland are left in a rain shadow. When the lows and their rotating air pass mainly over the western and northern land masses of Europe on their way here, they become much drier and their temperatures become more extreme than those moderated by the Atlantic, because the North Sea and the English Channel are too small to have much moderating influence.

The winds to which we are immediately subject are the following:

- Maritime Arctic and Maritime Polar, coming over the polar seas and ice in the north-east, north and north-west. Cold and damp.
- Maritime Temperate, from the north and south Atlantic. Damp, with temperatures moderate to warm.
- Maritime Tropical, from the south Atlantic over north Africa, Spain or southern France. Warm in winter, usually hot and dry in summer by the time it gets here. It is a relatively infrequent visitor to Britain but influential when it comes.
- Continental Polar, from north-west Russia. Very cold in winter, hot in summer, and again dry.
- Continental Arctic, passing over Scandinavia and Arctic ice. Usually cold or cool, and fairly dry.

Air flow at ground level is affected by friction between it and surface features. Then friction and disturbance diminish at higher levels, velocities increase, typically by as much as 50 per cent at heights of 200–300 m. If the wind is moving up a long slope, compression takes place between the upper strata of wind and those nearer the ground and this increases flow rates towards the top of the slope (see Figure 1.2). A valley may funnel wind energy still further and increase its velocity by restricting its release sideways.

In general, the greater the roughness of the topography, the greater the depth of the turbulence in the air passing over it. Uneven topography also causes flows to change into discrete, fast-moving parcels or

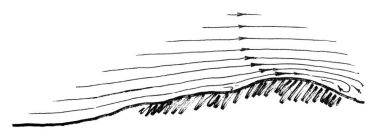

Figure 1.2
Compression and consequent acceleration of winds up a slope.

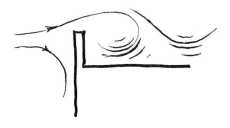

Figure 1.3

gusts of air, sometimes only 20 or 30 m across and lasting as little as 3 or 4 seconds. Because air is very elastic, the gusts form vigorously pulsing microvortices around and behind obstacles (see Figure 1.3). On buildings they typically develop just to the leeward of corners or over the edges of flat roofs or parapets. The wind velocities in vortices will be two or three times those of an undisturbed air stream. A particularly high risk occurs at the corners of flat-roofed buildings if the wind strikes at an angle that induces air flows to join forces over the two intersecting edges. Corners are therefore critical points in the design of wind-safe roofs and parapets (see Figure 1.4).

On a larger scale, gains in pressure on the windward side of a building will result in a loss of pressure to leeward and some cross-flow of air indoors, depending on how much leakage and filtration there is through the building envelope and the amount of indoor resistance to its cross-flow.

The wind effects of neighbouring buildings on one another often need to be understood, but because wind is invisible it is not easy to build up a systematic way of thinking about its likely behaviour just from observation and experience. Therefore wind tunnel tests are often commissioned when buildings have unusual shapes or relationships to others nearby.

Figure 1.5, based on a sketch by Lacy,[1] shows a simple example of an air flow pattern such as can be expected when wind flows among buildings. Air passing over the lower building would form horizontal, rolling vortices in the space between it and the higher building. Some of the energy in these vortices will gust sideways. The less disturbed air above the low building will exert some pressure directly against the higher building, but part of the energy will be deflected upwards, sideways and downwards. The relative amounts and therefore the relative air velocities across the facade will depend on the resistance and friction encountered in the different directions.

Downwards, the ground is an absolute resistance and energy will be deflected into sideways turbulence around the base of the building. Upwards and sideways, it will add energy to the air going around and over the building. Balconies, recesses and prominent architectural features will create de-energizing vortices on a facade and inhibit lateral flows, but will cause strong local concentrations. Any openings or cracks in the fabric will admit wind, and the less resistance it encounters in them the faster it will enter, sometimes carrying rain.

Two maps give useful information about the wind force distribution that typically results from our weather. Figure 1.6 (Lacy) shows the way average wind speeds in the zone from 10 to 250 m above open ground range from about 4 to 8 m/s. This has a bearing on wind cooling of the envelope and infiltration through it. The quietest areas in Britain are in the Thames Valley, the Midlands, the Scottish borders and the eastern half of the Scottish highlands, which all enjoy some shelter from the generally eastward wind flows.

However, Figure 1.7, also from Lacy, which shows the incidence of gales as days per annum, provides a strikingly different picture which is relevant for the safe design of buildings. Inland gales occur typically up to five days a year, rising to about ten days close to the coast, mostly within 20 km of it. Twenty to thirty days of gales occur along the western seaboard of England and the whole coast of Scotland, with an incidence of forty or fifty days on the outlying islands to the north-west and north of Scotland, and a similar range of incidence occurs in Eire. Gale force winds are a familiar cause of damage.

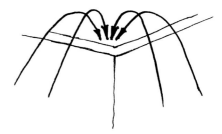

Figure 1.4

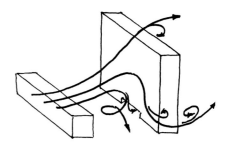

Figure 1.5
Typical wind pattern over and around buildings.

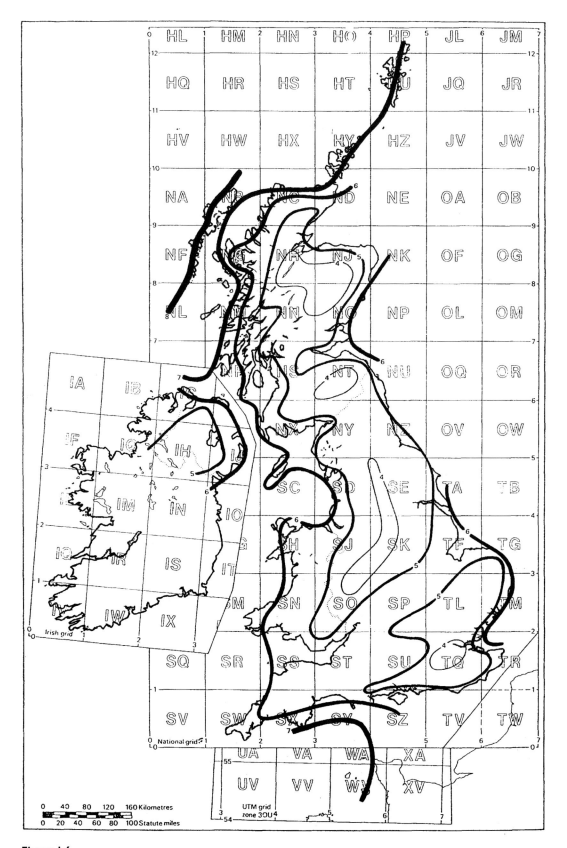

Figure 1.6

Annual average hourly mean wind speed (metres per second) in the United Kingdom at 10 m above ground in open country, based on data for 1965–9. Applies at altitudes up to 250 m above sea level (Based on Lacy, R. E., *Climate and Building in Britain*, HMSO, 1977, by permission of the Controller of HMSO: Crown copyright).

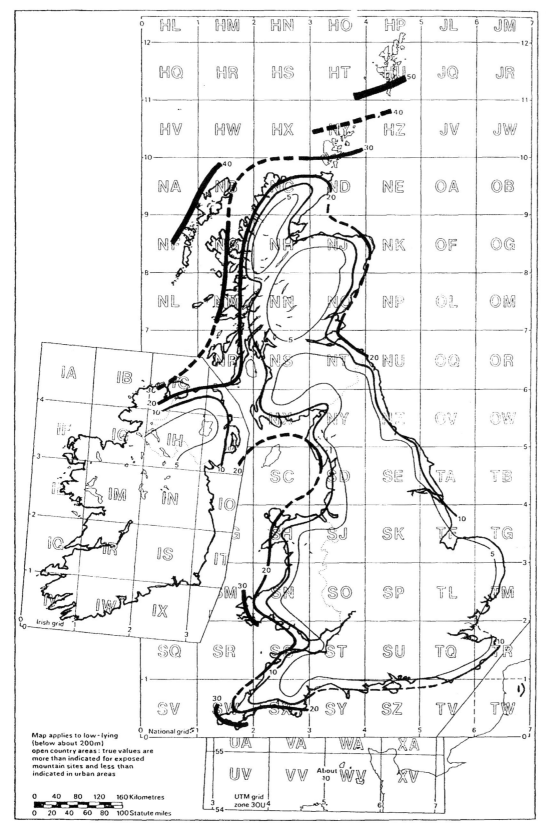

Figure 1.7
Average number of days with gale: (days during which the mean wind speed reaches or exceeds 17.2 m/s for at least 10 minutes),
1941–70 (Based on Lacy, R. E., *Climate and Building in Britain*, HMSO, 1977, by permission of the Controller of HMSO:
Crown copyright).

The best way to de-energize moderate wind near the ground is to make it do some work, for example by making it pass through a belt of trees where it has to move leaves and branches. However, experience in recent years has reminded us that at severe gale force velocities much damage can be done to large trees, especially in wet seasons when heavily foliated trees exert large overturning moments in soft ground, and when the trees go, whatever they were protecting gets full exposure.

1.4 Humidity in the atmosphere

Air comprises several gases mixed homogeneously and invisibly. Some are familiar, such as oxygen and nitrogen, but the pollutants mentioned below are also gases and so is water vapour. Each of them, by a process known as Dalton's Law, exerts its own partial pressure independently of the others, although their total pressure is what is measured as the lows and highs of weather systems, as described earlier. Nitrogen and oxygen are by far the largest components in the mix, whereas water vapour, although it plays a critical part in buildings and in our lives, generally makes up only some 8–15 mb (millibars) of the total.

The source of the water vapour is evaporation, mainly from oceans – especially, for these islands, the Atlantic. Here we have 200° azimuth of exposure to it from the Arctic to the Tropics and the generally easterly flows of our weather systems can then often carry large supplies of evaporated moisture, especially when the winds come from the south-west over warmer, and therefore more evaporative, water.

The amount of water vapour that can be held in the air depends mainly on the temperature of an air mass, although salts and dust in the air also have some influence. If the air is warm, its greater molecular activity enables it to carry more vapour than when it is cold, but when it reaches the limit for a given temperature it is said to be saturated. If it exceeds the limit, fine suspended condensate forms as mist or fog or shows as visible breath, for air exhaled from the lungs is highly humid. Liquid condensate also forms as dew on cool or cold surfaces. Indoors, steam is the result of the same mechanism, but at a much higher temperature. Dust helps to condense vapour, and the condensate that forms subsequently adsorbs some of the soluble contaminants which are suspended in the air.

The amount of vapour held in the air at any given time is described as its absolute humidity, and for thermal engineering design it may be evaluated either in terms of its partial pressure in the atmosphere or as its weight in kg/m^3. A measure is also often needed of the amount of vapour in the air relative to what could be held at any given temperature, i.e. at saturation, and this is termed its relative humidity (RH). This is technically its pressure at any given time expressed as a percentage of what it would be if the air were saturated at that temperature. For ordinary, everyday purposes it is sufficient to think of it simply as the percentage of vapour held in the air at some specific temperature.

People usually become aware of high humidities above RH values of 70 per cent or so, and at still higher values breathing feels less effective because the air becomes less efficient in removing lung moisture. The atmosphere is then commonly described as being 'close'.

Over these islands as a whole the absolute humidity in July is nearly twice that in January. The highest values generally occur in the south-west and the lowest in the north, both diminishing with distance from the sea. However, when air temperature is taken into account, the average relative humidity in January, even as measured in the relatively dry London area, lies between 80 per cent and 90 per cent throughout the 24 hours, while in July the day starts and ends at these levels but typically drops to 60 per cent or less by noon, when temperatures are high. Since temperature is the dominant influence and is, on average, higher inland than near the sea, the lowest RH values will typically be found inland in midsummer, specifically in England in a zone between Wiltshire and East Anglia and to the east of the Welsh hills. Close to the coasts, midday values typically stay higher, around 70 per cent, but the figure rises to 75 per cent or 80 per cent along the south-west seaboard, west Wales, west and north coasts of Scotland and the coasts of Ireland. They are probably typical of coastal zones of other areas in the world enjoying similar oceanic climates.

1.5 Rain

Figure 1.8, again from Lacy, depicts the average rainfall in Britain and this is one matter where averages are very relevant to design. Their relationship to the common weather flow eastward from the Atlantic is evident at once, and demonstrates the effect that high hills and mountains have in causing precipitation on the areas facing this weather as well as the shadowing given to leeward. Thus on the eastern side of western hills and mountains there is a sheltered zone running about 300 km northwards from the Thames which usually gets only 0.5–0.75 m of rain per annum, while the remaining areas get between 0.75 and 1.5 m, except for the mountainous zones, which receive upwards of 1.5 m and sometimes twice as much or more.

Steady rainfall results from cloudiness over large areas, which may be of the order of 1000 km across. Intense rainfall is usually produced by what are termed convection storms in which rain cools the column of air through which it passes, mainly by evaporation from raindrops, and thereby increases its density. At the same time the friction of the rain helps to create a downdraught that spreads as a squally outflow around and ahead of the base of a storm.

Prolonged rain occurs when there is little wind movement, so that most of the rain falls almost vertically. As a result, quite small projections, such as cills and eaves, provide more protection of facades than one might expect, and because prolonged rain can be so wetting, overhangs are particularly valuable in our kind of climate. The greatest benefits naturally occur where the Driving Rain Index is low. The higher the Index, the greater the overhangs need to be to give equal protection, or the more rigorously defensive the detailing of a building needs to be.

1.5.1 Contaminants in rain

Most of the world's encircling atmosphere is now polluted to some extent and in industrialized and heavily populated areas it can be assumed to be considerably or severely contaminated. Some of the contaminants get absorbed by water vapour and by rain, and examples of the damaging effects of this will be mentioned wherever they are relevant.

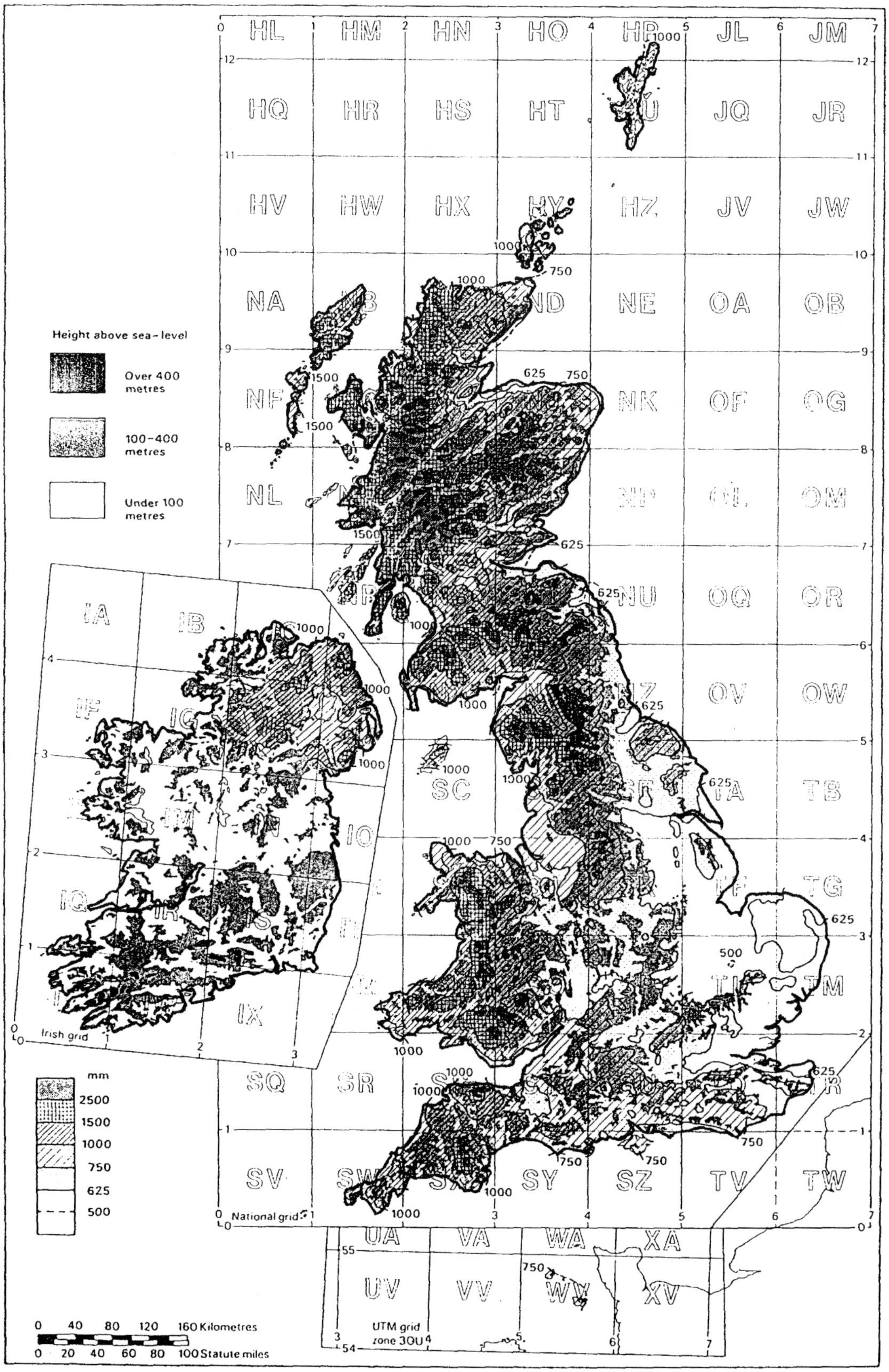

Figure 1.8

Average annual rainfall over the United Kingdom, 1916–50 (Based on Lacy, R. E., *Climate and Building in Britain*, HMSO, 1977, by permission of the Controller of HMSO: Crown copyright).

Volcanic activity and forest fires are unavoidable natural sources of contaminants. The manmade varieties derive mainly from the burning of fossil fuels, wastes of many kinds, some industrial processes, and internal combustion engines. Most of them end up in the air as carbon dioxide (CO_2), sulphur dioxide (SO_2), or nitrous oxide (N_2O) and are therefore acidic, although some industries produce alkaline gases. SO_2 has declined greatly since the 1950s, but the oxide of nitrogen has roughly doubled due to the increase in car usage.

The processes that produce contaminants mostly involve heat, so they habitually rise in the atmosphere; also, some industries are required to discharge them at a high level to avoid local concentration. Winds and thermals in dry weather disperse them into the air over a depth of a kilometre or more. Some contaminants rise higher and travel far, although in rain they deposit within 100 km or so, which means that in Britain appreciable or high concentrations should be expected in the vicinity of industrialized and heavily populated areas (see Figure 1.9). Their effects upon the materials of building envelopes will be discussed at relevant points throughout the book. A sea mist is neither rain nor vapour but a related state of water and it can be quite aggressive because of its contamination by sea salt.

Figure 1.9
Areas of substantial air pollution, mainly SO_2 at present.

1.5.2   Driving rain

Rain must be blown sideways if it is to make walls wet, and although the wind itself loses energy close to the wall surface, the momentum it gives to raindrops carries them forward against it and into cracks and crevices. Good anticipation of the severity of wind-driven rain can help determine what kinds of protection are needed in the locality for which a building is being designed.

An index-contoured map of driving rain severity was developed by Lacy in which the contours are the equal value products of the annual mean rainfall in metres and the annual mean wind-speeds in metres per second. Figure 1.10 shows the contours for our whole archipelago of islands in the form of the now familiar Driving Rain Indices.

The wind strengths used are those found at altitudes below 70 m in open country and at 10 m above local ground level. This therefore gives the designer no precise information about any particular site or project: the contours simply represent the amounts of rain that will be driven on to a vertical surface facing the wind over an average year. As Lacy says, it is not the absolute amounts that are useful so much as the evaluation of relative risks in different areas, which call for differing degrees of discretion in detailing and construction. He differentiates sheltered, moderate and severe conditions as follows:

- *Sheltered*   An Index value below 3, but in areas within about 8 km of the sea or a large estuary, exposure should not be taken as less than moderate.
- *Moderate*   Between Indices of 3 and 7, but any areas over a value of 5 within 8 km of the sea or a large estuary should be graded one step up to severe.
- *Severe*   Indices over 7 and including high buildings standing well above their surroundings.

Sites on slopes or tops of hills should be assumed to be one grade up.

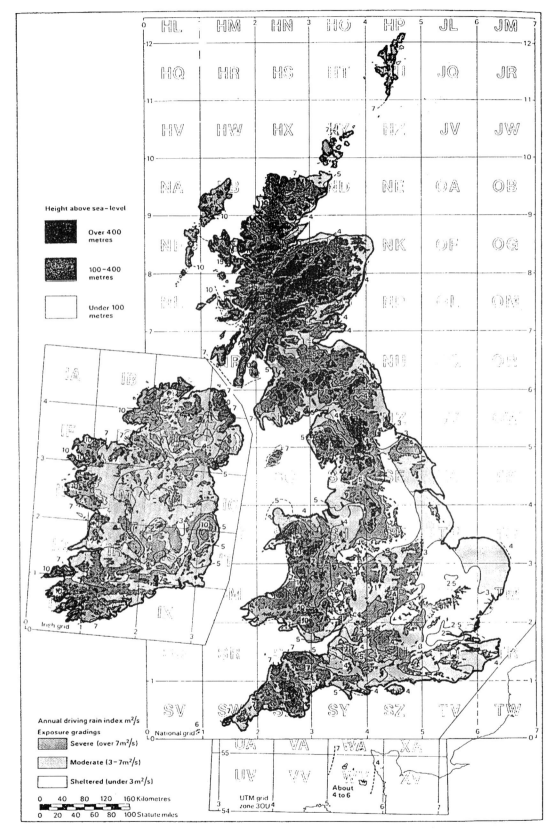

Figure 1.10
Map of annual mean driving-rain index for British Isles, in m²/sy. Data from Meteorological Office, Bracknell, and from Department of Transport and Power, Dublin. (Based on Lacy, R. E., *Climate and Building in Britain*, HMSO, 1977, by permission of the Controller of HMSO: Crown copyright).

Case note: Tredegar

Lacy illustrates the importance of making supplementary local evaluations by an astonishing example of comparative data from a case study in the town of Tredegar at the head of a valley in South Wales.

Tredegar is about 30 km from the sea at an altitude of about 300 m, rising to nearly 400 m in places. A group of houses was built at the upper level of the town on a west-facing slope about 700 m distant from the centre of the town some 90 m below. The houses suffered severely from rain penetration, condensation and some frost damage. Comparative data were taken from the housing site and from a site in the town centre, and it was found that the amounts of driving rain differed on average by no less than 6:1 and on some days by as much as 15:1. The mean wind-speeds differed in about the same ratios but the actual rainfall was about the same on the two sites.

Lacy also compared the Tredegar data with some from his base at the Building Research Establishment (BRE) at Watford, just north-west of London. The rainfall at Tredegar was about two and a half times greater than that at Watford with wind-speeds on rainy days about 50 per cent greater, but the total of driving rain was eleven times greater.

Tredegar is sited on an unusual configuration of land, but given that similar differences of height and exposure can easily occur between high- and low-rise buildings in urban areas, the case is a salutary reminder that the Driving Rain Index is only an initial guide to the relative severity of weather risk in different locations and with different heights of building.

There are not many data about the duration of driving rain, although Lacy mentions a few in his Tredegar study. The largest amounts of driving rain in continuous spells known in Britain were produced by storms lasting about 15 hours while the longest known continuous rain lasted 33 hours, though if some minor intermittency is ignored, longer periods have been recorded. Prolonged rain has an exceptional wetting effect on buildings because of the lack of intervals in which drying can take place or because capillaries and crevices get well filled.

For purposes of drainage design, the maximum rainfall rate in England, Wales and Scotland can be taken to be about 75 mm per hour, although rain falls at roughly twice this rate for about 4 minutes once in 50 years.

1.6   The mechanics of rain penetration

As remarked earlier, walls are wetted by the momentum given by wind to raindrops, and if this is sufficient, the water will get some way into holes and gaps. If there is through-passage for air, the water will be blown deeper into it by the wind, sometimes all the way through. However, other mechanisms are also at work. Capillarity and thermal pumping both exert some pulling power, and ordinary absorption is a familiar wetting process.

1.6.1   Capillarity

A crack or a hole with smooth, non-absorbent sides may act as a capillary, and by the mechanics of surface tension it will draw in low-viscosity liquids such as water. For this to happen, the cross-section of the crack or hole needs to be less than a millimetre, with optimum effect at around 0.1 mm.

In buildings, smooth, non-absorbent surfaces can be approximated by combinations of glass, plastics, metals, non-adhered mastics and even glazed tiling or mosaic where dense grouts are used and adhesion has failed. However, it is more common to find that the surfaces of gaps are uneven, dirty and to some degree porous. Capillarity is correspondingly reduced and instead there is some absorption into the materials themselves as the water goes in.

1.6.2   Thermal pumping

If a sealed container is warmed or cooled, the pressure of the air inside will rise and fall as its temperature goes up and down. Voids exist in numerous forms in the walls and roofs of buildings, sometimes by design (notably to contain insulation) and sometimes unintentionally. However, in building construction no void is likely to be perfectly sealed: it will be connected to the outdoor or indoor air by small leakage channels through which some of the air will be expelled or drawn in as the contained pressure rises or falls.

The air drawn in will always contain some water vapour, and if any of the walls of the void is close enough to the weather face it will often get cold enough to make this vapour condense. Any absorbent material contained in the enclosure or forming its walls will also become damp or wet, and because evaporation is very slow in such confined spaces, liquid water sometimes accumulates. Examples of both designed and inadvertent voids and likely types of leakage will be described and discussed where relevant throughout the book.

What is sometimes described as a 'breathing' action in the successive wetting and drying of concrete or brick surfaces seems to be a micro-porous behaviour similar in principle to this pumping action.

1.6.3   Surface absorption and evaporation

Facade materials usually range in absorption from moderate down to zero. Because intermittent wetting is a characteristic feature of the weather pattern of the British Isles, an absorbent facade has the considerable merit of holding some of the rain for later evaporation instead of letting it all run down as free water to test the vulnerability of capillaries, gaps, cracks and joints. Conventional facing brickwork is fairly absorbent, whereas curtain walling is not absorbent at all. Somewhere mid-way between them is engineering brickwork with dense mortar joints and external tiling or mosaic, as well as building stones for these have a range from moderate absorption down almost to zero. The less absorbent the facade materials, the more carefully the water-protective detailing needs to be thought out and built.

Wood can be coated to shed water, but paint coats are commonly broken at joints by thermo-moisture movement of the wood. Water or water vapour then has access to the body of the material, especially via its end grain, although its escape through the coating will be restricted.

Other absorbent solids take in and retain moisture by various physical and sometimes chemical mechanisms; the moisture retains some volatility as vapour moves in and out, depending on the relative vapour pressures in the adjacent air and the material as discussed later in

Chapter 2, where the indoor fabric of the building is examined as a vapour reservoir.

Moisture absorbed by materials raises their conductivity and thermal mass. The conductivity increases heat loss from a building and external drying increases it further as latent heat is removed by evaporative cooling. The drying rate increases as wind increases, and depending on the prevalence of winds in some regions, the cooling effect can become an influential factor on energy loss and indoor heat requirements.

1.6.4   Dirt adhesion

Surfaces which are both cold and damp allow dirt to be deposited and retained by physical and chemical mechanisms, and the relative persistence of these conditions in climates such as prevail in these islands, together with considerable dirty industrial discharge in some areas, explains the soiled appearance of many buildings. Exposed concrete is perhaps the saddest and most frustrating example in climates like that of these islands.

1.7   Snow

Depths of undrifted snow in Britain seldom exceed 12–15 cm on ground at low altitudes but in severe storms as much as 0.3–0.6 m or even 0.75 m have been known to fall over wide areas. Occasionally, as in 1881 in Hampshire and in 1963 in South Wales, as much as 1.5 m has been reported. It seems that the density of settled snow tends to rise initially and then stabilize at about 300 kg/m³. Data are limited, which is surprising in view of the long-established snow load allowances for roofs. This is something to watch in roof design in snow-risk areas.

1.8   Hail and thunderstorms

Most hail in Britain is small and harmless, but stones big enough to break glazed roofs fall in hailstorms about three times a year, mostly in midsummer and almost exclusively in southern England. About one year in five some very large stones fall – Lacy mentions a stone 80 mm across which fell near Cardiff in 1968.

Damage by lightning seems to be much more frequent, although reliable statistics are not available. Strikes to earth are believed to occur about once a year per square kilometre in most of England but much less frequently in much of Scotland, Wales and south-west England.

1.9   Solar heat gains and radiation losses

Heat from the sun arrives at the earth as direct radiation travelling line-of-sight, and as diffuse radiation from the atmosphere and clouds warmed by the sun. Heat is also radiated out into space at different wavelengths by everything on the surface of the earth, with loss of heat being greatest on cloudless nights. When cloud cover is available to act as a screen and insulator, this heat loss is much reduced.

Radiation losses and gains from these islands vary from place to place to a surpising degree for such small territories. The islands lie between the latitudes of 50° and 60°N with the result that the theoretical sun-day in summer is about two and a half hours longer in the north of Scotland than in the south of England, with more or less the reverse occurring in winter. Most of England, Wales and Ireland typically get 5 or 6 hours of bright sunshine daily in midsummer, with 7½ to 8 hours in a favoured strip along the eastern half of the south coast of England. The Pennines, the Scottish border, Scotland and Northern Ireland fall slightly below the 5- to 6-hour bracket, and in midwinter they all drop by as much as 75 or 80 per cent, especially in the mountainous areas of Scotland and Wales.

1.10 Weather effects on envelope surface materials

In the context of what has been said about rain, sunshine, air temperature, humidity and absorption, I offer some general comments on the way these affect the behaviour of the main types of external facing materials as background for the detailed discussions in later chapters.

1.10.1 Thermal and moisture movement

All materials change size when they are warmed or cooled. The amounts and rates will depend in abstract on their thermal movement coefficients which, in practice, will be modified by their thermal mass, i.e. their heat capacity. A product with a low thermal mass will respond to high or low temperatures in minutes or perhaps an hour or two, especially if it has thermal insulation behind or beneath it to prevent heat loss. A material or an element with a high thermal mass such as concrete or brickwork may take several hours, and sometimes so long that in a normal day or night it does not have enough time to respond fully to changing environmental temperatures.

If the body material of the product is a good insulator, the heated side will expand more, and more quickly, than the other side. The product will then try to bend, as is also the way with things made of two different materials intimately connected, and sometimes the bending becomes cumulative and permanent. Some types of marble do this if cut thin (see Chapter 4).

Products which can absorb moisture will also change size as they gain or lose it; again time is a factor, some materials being quick and others slow to do this. Bending also takes place if one side wets while the other remains dry.

The BRE drew the available moisture and temperature data together in an excellent trilogy of its monthly Digests (Nos 227–9, July to September 1979) from which I have constructed the detailed tables which form Appendix 1. Some general points from it are noted here.

1   The temperature ranges which building materials and elements typically experience *in situ* in Britain and Ireland are of the following orders:
    Clear glass 65°C
    Heat-absorbing solar control glass 115°C
    Metals (sheet and small extrusions) 75–105°C
    Plastics and bitumens 85–105°C
    Wood 105°C

Bricks, blocks, stone and concrete (the highest mass and therefore the slowest to respond) 70–85°C

2 Among the metals, copper, zinc, lead and aluminium have the highest thermal movement. Rigid plastics are higher still, a few much higher. Bituminous products have high coefficients but most are too weak to have effects similar to those of strong materials. The following are the thermal movements of some common metals, plastics and glass:

Movement mm/m per 10°C change of temperature
Copper 0.17
Zinc 0.30
Lead 0.29
Aluminium 0.24
Steel 0.12
Stainless steel, austenitic 0.18
Epoxy 0.45–0.55
Perspex 0.50–0.90
PVC and uPVC 0.40–0.70
Nylon 2.80
Glass: clear 0.7
Solar control 1.3

3 All the wet-manufactured products, and especially those bound by cement, shrink as their moisture content drops to long-term equilibrium. The process is quick at first, but the rate diminishes and can take several years to complete, depending mainly upon the thickness of the wet element. The original dimension is never recovered by subsequent moisture expansion, although a lesser degree of movement continues to take place as equilibrium conditions shift.

4 The dry-made products, chiefly the clays, come dead dry from the kiln and expand gradually while taking up moisture until they reach their equilibrium condition. Again it can take several years, and subsequently they too have only limited reversible moisture movement.

5 Timber products change size with their moisture content and also with the type of wood, the softwoods being greater and quicker in response than the hard. Among the softwoods, natural growth produces significantly less responsive timber than the quick growths of commercial forestry. Movements across the grain are much greater than along it.

6 The reflectance values of surface colours of materials have a substantial influence on their temperature gain and loss by radiation, and consequently on their daily and seasonal thermal movement.

7 Very important: the movement of moisture-absorbent materials is greatest between hot–wet and cold–dry conditions, and the amounts of movement are approximately additive for the two causes. It is convenient to think of them as thermo/moisture coefficients of movement. Product makers and test laboratories do not always provide both these key data nor do they often link them. Always try to get them.

8 Temperature conditions at the time of assembly on-site have a significant bearing on what happens afterwards to all products which are capable of exerting large and strong thermal movements. Materials may be as much as one millimetre per metre different in size between midwinter and midsummer fixing, and this can be a problem with, for example, pre-punched products and metal trim assembled end-to-end.

9   Moisture-sensitive materials make their most extreme dimensional changes in periods of prolonged wet or dry weather.

The cumulative effect of all these thermal and moisture movements in this island climate is to create an unusual degree of restlessness among the various products and materials making up the envelopes of buildings. It causes damaging collisions or opens up gaps and cracks by which rain gains entry to the fabric, or air and vapour reach or leave the interior of the envelope and the building more easily.

1.11   Summary: a hostile climate

The climates of Britain and Ireland are hostile to the building envelope, and the following are the chief reasons:

* The exceptional frequency of freeze/thaw cycling when envelopes are wet
* The occasional prolonged changes from damp oceanic to dry Continental-style weather
* Generally slow drying rates
* The frequent dimensional changes of absorbent materials from wet and hot to cold and dry
* The frequent association of strong winds and heavy rain
* The lowering of insulation values by high and pervasive humidity
* The combination of high atmospheric humidity and thermal pumping
* The combination of high humidity and atmospheric contaminants.

The immediate cause of our problems is a matter of geography. The islands are exposed on one side to the wetting influence of the Atlantic in the troublesome freeze/thaw temperature range, and on the other to drier Continental weather with more extreme temperatures. The location of the islands ensures that we experience both types everywhere.

Prolonged dampness encourages chemical reactions that cause degradation of materials, the growth of algae, lichens and moss, the accretion of dirt, and corrosion. This largely explains the discouraging appearance of concrete, renderings, and some other forms of cladding in Britain compared with the relatively clean buildings that enjoy Continental climates.

Most buildings are designed for a life expectancy greater than 50 years, sometimes much greater, and accordingly need to be detailed to withstand the weather severities to be expected within the intended time span. One always has to work with the climate; fighting it is a losing game.

Reference

1   Lacy, R. E., *Climate and Building in Britain*, HMSO, 1977. Embodies many years of experience by its author, who was the climatologist at the Building Research Establishment.

Further reading

1   Chandler, T. J. and Gregory, S., *The Climate of the British Isles*, Longman, 1976. A valuable general reference on climate.
2   Hanwell, J., *Atmospheric Processes*, Allen & Unwin. Excellent descriptions at a fundamental level of the way the atmosphere functions, at global scale, local UK scale and indoor comfort scale.

The following BRE Digests deal with wind:

No. 119, July 1970: The Assessment of Wind Loads.
No. 141, May 1972: Wind Environment Around Tall Buildings.
No. 283, March 1984, The Assessment of Wind Speed over Topography.

In respect of wind and rain together: No. 127, March 1971: An Index of Exposure to Driving Rain.
4   For a fundamental engineering discussion of wind behaviour around buildings, see Cook, N. J., *The Designer's Guide to Wind Loading of Building Structures*, Butterworth, 1985. The book comes from the Building Research Establishment.
5   For an interesting diversion from the mainstream building science with which we are familiar in Britain, see Hutcheon, N. B. and Handegord, G.O., *Building Science for a Cold Climate*, Wiley. The authors were, respectively, Director of the Building Research Division of the Canadian National Research Council, and a member of the staff of that organization. The discussion of climate, indoors and out, is relevant and good.

2   The indoor climate

2.1   Introduction

Indoors, the designer's most natural need is to provide for personal
comfort. For most people this lies within an air temperature range of
19–21° or 22°C, with surrounding surface temperatures not much below
or above these figures – ideally a little above – and relative humidity
(RH) values somewhere between 25 per cent and 70 per cent. People are
not particularly sensitive about relative humidity, but persistent levels
below this range cause dry skin, sparking and other minor nuisances,
while above it clothing and bedding soon become noticeably damp,
feelings of 'closeness' develop, respiratory discomfort occurs, and
sometimes ailments.

Buildings and their contents are in some ways more sensitive than
people. Persistently dry conditions cause doors, wood trim, furniture,
floor joists, boarding, plasterboard and some types of carpet to shrink,
while prevalently high humidity causes enlargement, tightness, and
sometimes damage. Changes between low and high, which are
troublesome enough in ordinary buildings, are unacceptable in such
places as museums or art galleries where conservation of organics
requires thermo-moisture stability above all else, or in places such as
cold stores, computer rooms and certain kinds of laboratories where
comfort is irrelevant, but stable and specific indoor climates are necessary.

We saw in Chapter 1 that the oceanic climate of Britain and Ireland
commonly results in naturally high absolute humidities in the open air.
The residual moisture content of indoor air is basically of the same order
but to this is added the moisture produced by people and their activities, and this can quite quickly build up to levels at which there is a risk
of condensation and discomfort, sometimes ill health if prolonged. The
added moisture must therefore be enabled to escape through the
envelope or be removed to atmosphere by power at a rate roughly equal
to that of its creation.

That is usually the key point in designing for humid climates: the
indoor vapour pressure must be able to drop towards the outdoor level
reasonably quickly, e.g. within half a day or a day. By contrast, in dry
climate regions vapour often needs to be retained and the aim then
becomes a vapour-tight envelope.

In order to design logically for good vapour management we need to
understand the mechanics of vapour behaviour indoors, the amounts

involved, the interaction of vapour with temperature and the mechanisms by which it can pass through the envelope or can be stopped, and then we need to relate this information to the kind of building being designed and any special requirements it may have.

2.2 Sources of indoor water vapour

When a space is newly enclosed, construction moisture begins to contribute to indoor moisture by evaporation and with wet construction this lasts a year or two. Then, when the space begins to be used, the other sources begin to contribute, which are:

- Breath, mainly human, sometimes animal. This has an RH close to 100 per cent and is usually the biggest contributor
- Cooking
- Combustion moisture from gas or oil used for cooking, heating (by flueless burners) or industrial processes
- Evaporation from washing and drying activities
- Evaporation from swimming pools and water-based leisure facilities
- Wet industrial processes
- Transpiration from planting
- Humidification by air conditioning

Quantities can be attached to most of these sources and are given in Table 2.1 as equivalent amounts of liquid water in litres. One litre weighs 1 kilogram (kg) and is approximately 1.76 imperial pints.

Table 2.1 Water vapour production

Source	Quantity of water (litres)
Human breath	
Resting or asleep	0.04 per/h/person
Sedentary	0.05 per/h/person
Heavy work, sweating	Upwards of 0.1 per/h/person
Family cooking (but excluding the vapour from gas if this is the cooking fuel)	0.9–3.00 per day
Combustion (flueless appliances)	
Gas (typical household cooking)	0.65 per hour
Gas, basic rate, heaters	0.80 per m^3
Paraffin heaters	1.00 per litre of fuel
Household washing and laundering	
Clothes washing indoors	0.5–1.8 per day
Clothes drying indoors	5.0–14.0 per day
Bath, showers, dishwashing, (per person per household)	0.75–1.5 per day
Planting	
Large pot plant, watering and transpiration	Up to 0.85 per day

Industrial processes, pools, leisure facilities based on water, mass planting, etc., are specialist items, normally dealt with on an individual basis. Milbank[1] has provided a formula for estimating moisture emission from water surfaces.

The amount of construction moisture depends on how much water gets built in. As a rough guide, a typical 100 m² house built with cavity brick or block walls, concrete ground floor and pitched tiled roof may contain 3000–4000 litres of water initially. A fully exposed house should evaporate 50 per cent or more direct to atmosphere, but in terraced housing less can escape this way because the external surface per dwelling is not as great. In flats with concrete party floors there will be more moisture and still less external surface. Evaporation to adjacent houses or flats will not work because they, too, are trying to lose moisture to their neighbours. As a result, indoor evaporation will enhance indoor humidity for longer periods.

Animal breath is chiefly important in relation to zoos or other animal accommodation.

2.3   Indoor vapour diffusion and escape

2.3.1   Vapour diffusion

As noted in the discussion on the outdoor climate, water vapour is a gas which exerts pressure, and when it is produced in an enclosed space, its pressure diffuses it throughout the space as an addition to the vapour already present so that in a few minutes there is a distributed increase in the pressure level. If the boundaries of the space can absorb moisture and if there are soft furnishings and other absorbent items in the room, some of the increased humidity will enter these and transient storage will take place. If there are ducts or openings to other spaces, the vapour will diffuse into them until the pressure differentials equalize.

As the pressure indoors becomes higher than the pressure outdoors, it will attempt to escape, partly by migration through any of the envelope solids that are permeable, partly by leakage through the small free air passages that are always present, or by chimneys, and partly by any air changes resulting from incidental or positive venti-lation. When the indoor and outdoor pressures are again equal, the amount of water vapour remaining indoors will be at a natural minimum for the time being and any further reduction will require some form of dehumidification. This will then cause the outdoor vapour to have the higher pressure, and it will begin to enter by the same routes that the indoor vapour used for escape, once again trying to reach equilibrium.

Vapour production indoors obviously rises and falls. Some envelopes restrict vapour migration more than others, and the capacity for transient storage goes up and down. If over a period of time the rate of indoor vapour production were to exceed the combined rates of vapour escape and storage, a state of saturation would eventually be reached and condensate would form not only on windows but on walls and floors and within the building fabric and the furnishings. If the imbalance is marginal and there is a short spell of dry weather, conditions indoors will improve, but if there is a considerable imbalance, problems can become cumulative. This is what happened in thousands of British and Irish dwellings in the late 1960s. Cases are discussed later when we look at specific building types.

In air-conditioned buildings the problem is reversed. The vapour content can be kept sensibly constant but only if its movement through the envelope is much restricted or prevented so that gains or losses by these routes do not become significant.

2.3.2   Rates of dispersal

It is difficult to appreciate the rate at which water vapour moves about indoors because it is invisible, it can rarely be felt and it has no odour. However, its behaviour resembles odours, for these are gases like water vapour and behave very much as vapour does, but with the advantage that we can readily sense them by smell. So, by imagining the diffusion and dispersal of water vapour as if it were an odour, we can get a fair idea of the rate and extent of its distribution in a building. Naturally, the greater the pressure differentials between one place and another, the more rapid the flow will be from the higher to the lower, whether it is water vapour or an odour. However, the pressure differentials are small, so the movements are slow; there is no noise from them passing through a gap as there might be from wind.

2.3.3   Transient storage and reservoirs

The moisture absorbed by furnishings, clothing and permeable building materials is volatile. It interchanges with the air by moving back and forth according to the relative temperatures and vapour pressures in the air and the reservoir and the ease with which the moisture can move in and out. A reduction of the temperature in a material or an increase in the vapour pressure on it will cause it to become damper, and vice versa.

Clothing and soft furnishings generally receive and give up moisture easily. Foam plastics are slow. Books in quantity are slowish, as are plaster and softwoods, and these latter become slower still when coated by resin-based paints, or by plastic substitutes for wallpaper, both of which restrict the passage of moisture. Modern paints are mostly resin-based and the resin is relatively impermeable, but paints are now usually microporous and more permeable. Nevertheless, the great increase in the use of low-permeability coatings, building materials and furnishings has reduced the reservoir uptake rate and capacity in modern buildings and is one cause of their increased sensitivity to condensation.

These reservoirs are now known to be much more important than was formerly thought, but they have had little attention in research.

2.3.4   Vapour passage through the envelope

In wintry conditions when doors and windows may not often be open, vapour uses leaks and permeable solids as movement routes outwards or inwards, depending on the balance of the indoor/outdoor pressures.

The leaks are typically small gaps around doors and the opening lights of windows, loose-fitting frames, air or fan vents, loose laps in timber or metal cladding, and faulty seals around pipes. In still air, the pressure differentials will try to equalize through them, but if air as a whole is attempting to pass through, i.e. when there is some wind, it will usually take charge by acting as a vapour carrier.

All permeable solids acquire a moisture content more or less in equilibrium with the air around them, but when pressure builds up on one side the molecular moisture moves into the solid on that side, eventually increasing the moisture content throughout so that the evaporation rate on the other side increases. This is the migration process, and its rate depends mainly on the pressure differentials, the permeability of the solids, the length of the path through them and the resistance offered by decorative finishes. Ventilated cavities in walls and ventilated roof lofts provide short-cuts for vapour escape. The permeable solids of envelopes are materials such as plaster, mortar, lightweight blocks, brickwork, and timber. Dense concrete restricts permeability, and so does plywood, because of the glue laminae. Glass, metals, mastics, plastics, bitumen felts and asphalt are for most practical purposes impermeable.

2.3.5   Troublesome effects of insulation

Modern requirements for increased insulation of the building envelope, introduced in and after the 1950s, have had some unexpected effects. Additional insulation had seldom been thought worth while in the temperate climates of Britain and Ireland, so that even in cold weather the escaping heat could usually keep the whole envelope above dew point temperatures.

The introduction of insulation into envelope systems has altered this by limiting the outflow of heat so that the outer parts of envelopes often become cold enough to cause condensate to form on or in them, sufficient even to run a risk of frost damage. Sometimes condensate forms in the insulation itself. The logical consequence is the use of barriers to prevent indoor vapour getting into and beyond the insulant, but what has then often been overlooked is that by this means the outflow of indoor vapour is prevented, although the condensation risk on indoor surfaces may not be much increased because the insulation keeps them warmer and thus raises their moisture-carrying capacity.

2.3.6   Passive ventilation

Two natural ventilation mechanisms, stack power and wind, are at work in all buildings and help to remove excess indoor vapour. They need to be understood.

Stack power

A temperature gradient naturally develops in any body of air so that the upper parts become warmer than the lower. Power is latent in this situation and comes into operation if the warmer air can escape at an upper level so that replacement air can enter lower down (Figure 2.1). The larger the temperature gradient and the higher the stack (i.e. the greater the vertical distance between inlets and outlets), the greater the power will be. It operates more vigorously therefore in two- or three-storey dwellings than in flats or single-storey houses. In high-rise buildings, stacks can be tall and strong and need horizontal interruption, usually at each storey. It is a significant factor in air-conditioning design. This storey-to-storey disconnection serves an important second purpose in preventing the upward movement of fire, for fires create fierce stack power.

Some storey-to-storey leakage always survives so that lower storeys have some negative pressure while upper storeys become positive. At some point there will therefore be a neutral plane and its location is a factor in air-conditioning design and operation for medium- and high-rise building.

Stack power also increases air leakage through the upper envelope. Brand[7] says that in a ten-storey building it can be powerful enough to generate as much pressure through a wall as does wind (about 0.13 in. of water). He further concluded from a building he studied that for every 10 m^2 of conventional cavity masonry walling, the air leakage could carry with it up to 0.5 kg of water in vapour form in addition to normal migration. This is quite a lot.

When the entries and exits for air occur at various levels in an envelope the stack height and power will be diffuse, but when provision for it is made deliberately by properly placed openings, the relative ease of air movement makes it more specific and it becomes calculable as a ventilation mechanism.[2]

When fires occur, the intense stack power and air pressure increase are mechanisms by which toxic smoke diffuses in a building, finding all sorts of channels. It is deadly.

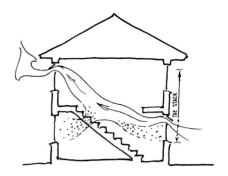

Figure 2.1
Stack power via openings and leaks.

Ventilation by wind

In principle, air changes can be induced by wind, which creates positive pressures on one side of a building and negative on the other. As a result, a push–pull circuit can develop across the building by infiltration on one side and exfiltration on the other unless intermediate subdivisions of space wholly prevent it. Excess indoor vapour will then merge with the vapour in the air passing through and be carried away with it as it departs.

Fireplaces

Open fireplace flues provide continuously available free air paths to atmosphere. Vapour pressure differentials are therefore able to equalize easily, and there is also a more or less continuous upward air flow, so that flue draughts become effective vapour carriers. Air inflow to balance the outflow will depend upon air leakage and the occasional opening of doors and windows, or there might be specific provision for the admission of fresh air.

The natural power of flues results from a combination of stack effect inside them due to the normal warmth of a chimney in a warmed building, and to what might be called an atomizer effect. This is when wind blows over the top of a flue, causing a reduction of pressure at the top which in turn draws the flue air upwards. A fire in a fireplace naturally accelerates the upflow, and if its flue is one of several clustered in one chimney they will all be accelerated by the increased warmth. This largely explains why pre-1939 British houses enjoyed freedom from condensation, being usually equipped with several fireplaces. Even nowadays it is unusual to have troublesome condensation in any dwelling with an open chimney flue.

Not only traditional fireplace flues provide this valuable form of ventilation. It is now realized that vertical ducts for natural ventilation can be introduced in various ways into buildings that lack artificially powered air changing, and further reference will be made to these in the later discussion of building types.

2.3.7 Simple powered aids for moisture removal

Extractor fans

In oceanic climates, extractor fans are commonly fitted in kitchens to remove vapour as well as cooking odours at source, operating independently or in conjunction with cooker installations. The opening for the fan allows water vapour to move through continuously unless it is closely shuttered; the fan merely ensures and accelerates a positive outflow.

Fans are usually operated for only short periods when this is a personal decision, but when controlled by a humidistat they are known to run many times longer, confirming that people find it difficult to judge levels of humidity.

Dehumidifiers

Dehumidifiers draw indoor air into contact with a surface chilled below dew point so that water vapour will condense on it. The condensate then has to be drained away. If dehumidifiers are used, it is necessary to reduce or prevent the entry of outdoor air and its vapour; otherwise one is simply trying to dehumidify the outer atmosphere, which is impossible.

Case note

This concerns a large historic house open to the public in summer. In winter, it had minimal heat input: one big room was exposed on three sides and often became cold and damp. A dehumidifier was put in to run continuously and produced as much as 30 or 35 litres of condensate per week. This far exceeded any conceivable vapour entry via the walls and windows, but a large fireplace had a correspondingly large open flue which provided easy access for the inexhaustible supply of vapour from atmosphere.

Heat recovery extractors

Powered extractors which recover heat from the air they expel and return it for re-use have been developed for domestic use and are often associated with tubed flue systems in dwellings without fireplaces. Because such equipment can conserve energy at reasonable installation and running costs, it will doubtless find a useful place in design and specification.

2.4   Air conditioning

Air conditioning is needed in buildings where natural ventilation cannot provide a comfortable and sufficiently controllable supply of fresh air. It may also be needed where windows must be closed to mitigate external noise, or where internal heat gains are high, or where an indoor climate has to be thermally and moisturally stable at specific levels, for example in museums, archive stores, large computer installations and some research laboratories.

The design of air conditioning is a matter for engineering professionals, and only a few comments are appropriate here.

The accurate prediction of indoor climate should be possible where full air conditioning is installed, i.e. where air change rates, RH levels and temperature are controlled, and filtering is used where pollution-free air is needed. For reasons of hygiene it is usual to design the systems so that indoor air pressure is slightly positive to ensure that any envelope leakage is outward. However, reasonable air-tightness is increasingly valued to reduce energy waste, and a well-sealed envelope is essential when air conditioning has to perform to a tight specification. Architect and engineer should then agree the envelope detailing for the purpose, and the performance of the system should later be checked by independent commissioning before handover.

In a climate in which the average outdoor January temperature is likely to be above 1°C or 2°C and where conditioned air is required to keep indoor RH values below about 55 per cent, vapour barriers may not be necessary to protect insulants in the envelope. This is because indoor vapour will be removed at a rate which should ensure that any condensation will be limited, transient and soon evaporated.

There are some further references to air-conditioning design when we discuss indoor climates of specific building types.

2.5   Refrigerated spaces

When buildings such as cold stores or enclosed spaces such as laboratories have to maintain very low RH values, implying very low vapour pressures, the problem is to prevent external vapour from getting in through the envelope. This is the reverse of the normal situation. A vapour barrier system then has to be provided on the outer boundary of any void containing insulation rather than on the inner.

2.6   Relative humidity and indoor condensation

As explained in Chapter 1, relative humidity (RH) is the amount of water vapour held in the air expressed as a percentage of the maximum which the air can hold at that temperature. At its maximum, i.e. when the RH is 100 per cent, the air is said to be saturated, so that if the temperature were to drop, some of the vapour would have to condense as airborne mist or as liquid on any accessible surface below dew point.

When warmth and humidity levels are maintained within comfort limits everywhere in a building, troublesome condensation should not occur on indoor surfaces. However, if some rooms are unheated or poorly heated and yet can be reached by water vapour available from elsewhere in the building, then in cold weather the risk of condensation in them increases and may materialize. This is because the RH values will rise in the colder air and the inner surfaces of the envelope will be disposed to drop towards and perhaps reach dew point temperature.

The relationships can be made specific by reference to a diagram called a psychometric chart (Figure 2.2) which, in its simplest use, relates temperatures to RH values. Here is an example to illustrate this.

Example

Assume an RH value of 55 per cent and a temperature of 18°C for the main body of air in a building. This will be represented by point A on the chart. If it is then assumed that in an unheated room at the same vapour pressure the temperature drops to, say, 13°C, the RH value will be found by reading across horizontally to intersect the 13°C vertical (point B on the chart). The RH value is then seen to have risen to 75 per cent. If the temperature drops further to 10°C, the RH would reach 90 per cent (position C), and if the horizontal is projected further to intersect the 100 per cent RH line, the temperature for this condition is seen to be 9°C (point D). If therefore the weather is cold enough to cause the air temperature in a poorly heated room to drop to 12° or 13°C, poorly

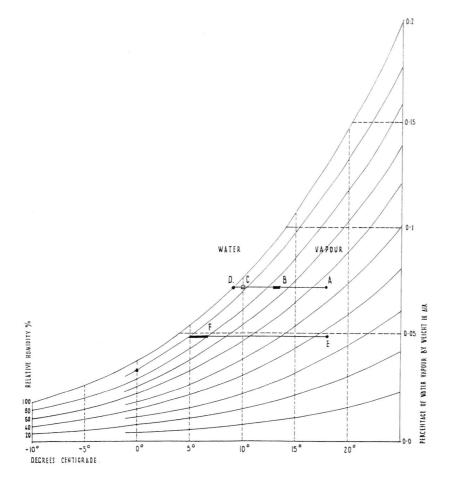

Figure 2.2
Chart relating moisture content and temperature.

insulated walls or the glass in a window or on a cold-bridge surface might well drop down to 9°C or less and surface condensation would take place because the air layer in contact would be at 100 per cent RH, i.e. saturation. If this were to happen frequently or persistently, vapour would condense on the surface and migrate into the envelope, dampness would become pervasive in it, and its natural insulation would be reduced.

Condensation can also occur behind furniture large enough to shield an area of wall surface from room warmth, and in corners – especially three-way corners at floor or ceiling level – where the high proportion of external surface increases heat loss and lowers the local air temperature.

Centrally heated (but not air-conditioned) buildings in a climate such as that of Britain and Ireland, running at temperatures of about 18–22°C, usually drop naturally to RH values of 30 per cent or 40 per cent in winter and rise in summer to outdoor norms sometimes in excess of 65 or 70 per cent. The lower winter levels typically cause shrinkage cracking of the building fabric and softwood trim and joinery. Plasterboard will try to open up at joints; windows and doors made of low-grade fast-grown softwood become noticeably smaller and looser in their frames and joints open up. Conversely, at the higher RH values in summer, cracks close up, softwood swells, and windows and doors tighten up, often to the detriment of their hardware when they are forced open or closed. This caused rapid deterioration of many doors and windows in Britain during the 1970s and early 1980s.

In buildings that are not centrally heated, humidities in winter will generally be higher than in summer, the reverse of the situation just described.

Some useful information about the efficiency of ventilation air can be found in Figure 2.2. As noted previously, winter air in these islands often has high RH values, 85 per cent or more, at typical temperatures of around 5–7°C (point F) and although a rise of temperature indoors to 18°C will lower the RH to 35 per cent or 40 per cent (Point E) and increase the capacity of the air to carry away indoor vapour as changes take place, it will be nothing like as efficient as the colder, drier air of such climates as those of mid-continental Europe or North America. This partly explains why we sometimes need fairly high air change rates in winter.

2.7 Cold bridges

Cold bridges occur at positions in the envelope where the insulation is locally very poor and conductivity therefore high. They are usually found where good heat-conducting elements short-circuit across generally good resistance in the envelope, for example when solid construction is formed around doors or windows in an otherwise insulated cavity wall or where concrete wall framing is exposed inside and out. They act as condensers when exposed to unfavourable combinations of high RH values and low air temperatures indoors. In effect, cold bridges are messy dehumidifiers.

As a condensation nuisance, cold bridges have received exaggerated publicity. The worst cold bridge in almost any envelope is single glazing, and if condensation occurs persistently on it, this simply means that indoor RH values are running higher than they should. The necessity which developed in the 1960s to provide a catchment and drainage arrangement for window condensate simply reflected the failure at that time to understand the indoor humidity problem and to sidestep it in design; such arrangements should never have been necessary.

A particular characteristic of cold bridges is that, because of the preferential deposition of molecular dirt on cold damp surfaces, these darken more quickly than the warmer, better insulated and molecularly more active areas around them. This produces dark pattern staining, often resulting in an expensive need to redecorate frequently. If the bridges are big and numerous they can transport undesirably large amounts of heat out of the building.

In Britain, the traditional 9 in. solid brick house walls of the 1920s were entirely cold bridges by modern standards and the dwellings were often uncomfortably chilly, but because of their numerous fireplaces, the natural ventilation rates were too high for much condensation to occur. On the other hand, they did suffer from relatively quick dirtying of the walls because of their coldness and they needed frequent redecoration.

The reduction in the rate of redecoration needed for insulated envelopes is often overlooked: it is a significant economic offset to the cost of insulation.

Case note

The walls of a large up-market block of flats built before 1939 had walls of concrete cast *in situ* against an inner lining of cork squares. Their joints had not been taped at the back and concrete grout often ran through, forming a pattern of square linear or dot-and-dash cold bridges which darkened in a couple of seasons. The landlord was responsible for redecoration, and the demand for it was so high that most of his profits were eaten up by it.

2.8 Mould growth

Moulds grow readily where there are persistently high RH values, in excess of about 70 per cent. Spores are as widespread in the atmosphere as radio waves. Brundrett[3] says they enter most homes at a rate of about a million an hour and all they need for growth is moisture, food and quiet air. Moisture can be provided by condensate or by dampness due to leakage, and food can be airborne as volatiles from cooking, especially frying, and sometimes from the constituents of wallpaper and its adhesive. Moulds can be killed by proprietary washes or by bleach, but this is only temporary. For a cure, the cause needs to be removed.[4]

2.9 Dry rot

The fungus that causes dry rot is *Merulius lacrymans*.[5] Its spores are widespread, but it will only establish itself where it can feed upon wood or other cellulosic material with a moisture content in excess of 20 per cent. It will spread in a stable atmosphere of damp, cool, still air, and its tendrils can extend several metres if a food supply remains available. The nutrients are transported along the fibres. It cannot colonize dry wood, although well-developed strands can carry some water to moisten wood which would otherwise have stayed too dry. This weeping effect explains the descriptor *lacrymans*.

In appearance the fungus has silky fibres which form soft cushions and sheets of a silvery grey colour, tinged yellow and lilac. Dry rot has a distinctively mouldy smell by which it is often first detected.

High-risk locations are where ventilation is poor and humidities are high, for example among roof timbers, especially near the eaves where leaks often make a contribution to dampness, and behind panelling where the void is damp and the air is quiet. Damp brickwork or plaster is conducive to its development and it can spread across them and through any available crevices.

Wood affected by dry rot loses weight and becomes brownish, weak and crumbly, and it splits microcubically as it shrinks. As with moulds, if the wood providing the food source dries below 20 per cent moisture content, the fungus will slowly die. One form of treatment therefore is to starve the wood of moisture.

2.10   Diagnosing high humidity and condensation

There are several ways of distinguishing high humidities and condensation from leakage. The first is by feelings and odour. If the RH is really high, i.e. above 70 per cent or so, a feeling of closeness develops and breathing becomes restricted. There may be a musty smell generated by damp furnishings, clothing and the building fabric, especially if moulds are forming.

Hand-held instruments can be used to measure RH levels. Some need to be in the test atmosphere for a few minutes before they give acceptably accurate readings. Wet and dry bulb whirling hygrometers can be used and long-term recorders are readily available on hire.

The second is by visual evidence. It is usual to find that condensation shades off gradually from damp to dry on indoor surfaces while leakage stains typically have sharp edges, often with dried salts dissolved out by moisture passing through the materials. Also, condensation tends to be systematic, whereas leakage is more likely to occur related to rainfall. The time and circumstances of each occurrence should be recorded if anaylsis is attempted.

It is not uncommon to find both condensation and leakage in the same situation and it may be difficult to separate them; treatment may have to be undertaken in stages to determine and eliminate causes. Sometimes there is enough condensation in a concealed position to drip and give the appearance of a leak.

If moisture dripping from the ceiling under a flat roof is brown, it is very likely to be condensate. Moisture vapour from indoors may have passed through mortar joints in brickwork or have been in persistent contact with plaster or concrete, thereby acquiring alkalinity. The condensate is then able to react with low-grade tarry constituents in a vapour barrier or insulation or sheathing felt under asphalt, and it then becomes brown. Rain alone is usually acidic and clear. This evidence can help diagnosis.

2.11   Other climates

Other types of climate offer other indoor/outdoor differentials of temperature and vapour pressure and result in different moisture control problems. One instructive comparison for designers working in temperate oceanic climates is with mid-continental conditions, e.g. in central Europe or mid-west North America. The air in these places is basically drier, being far from large bodies of water, and winter air temperatures drop as low as −30 or −40°C and sometimes lower, too cold to hold much moisture. The air then becomes a very effective drying agent, and especially when raised to comfortable indoor temperatures. At the same time the indoor/outdoor vapour pressure differentials will be high and will try vigorously to induce moisture outflow so that, as noted earlier, the problem becomes the design of envelopes to restrict the escape of indoor moisture, the reverse of temperate oceanic needs. All this is reflected in the fact that Brand[7] in Canada regards an indoor RH of 50 per cent as high, while in Britain and Ireland it is low to moderate.

Another climate comparison can be made with tropical oceanic climates. In these, air conditioning is designed to reduce both the temperature and the vapour pressure (rather like the design problem with cold laboratories in Britain). The envelope then needs to exclude the higher external vapour pressure.

So it is risky to do a straight borrow for envelope design from other climates; it must always be thought through properly for the climate concerned.

2.12   Computations and judgement: design and indoor climate

It should now be clear that in a high-humidity climate a distinction has to be made between the design of envelopes for buildings in which vapour content is to be mechanically controlled as part of air conditioning and those in which it will be left to natural forces. The former need a vapour-sealed envelope so that computations of airborne moisture can be made reliably, while the latter need enclosures through which indoor vapour can escape naturally at a reasonable rate.

A really good vapour seal needs both a general barrier and the stoppage of leaks. The latter is more easily said than done and needs both positive detailing and careful site inspection; but good approximations can be achieved.

Natural vapour escape is another matter, for the database is very poor and reliance has to be placed mostly on judgement. Designers will be helped to develop the necessary rationale by the detailed discussion of walls and roofs which takes place in later sections of this book, but there are some larger parameters which can best be addressed now by looking at a few selected building types.

2.13   Designing indoor climates

2.13.1   Residential buildings

From Victorian fresh air to modern condensation: over the 250 years or so up to the mid-twentieth century the heating of dwellings in Britain slowly shifted from the burning of wood in medium and large fireplaces to the use of coal in small grates. Most of the earlier fireplaces were reduced in size and sometimes reduced again, incidentally providing interesting opportunities for historical rediscovery in the second half of the twentieth century. Central heating by hot water radiator systems was introduced into housing on these islands from the late nineteenth century onwards, but not very rapidly because the climate was often not thought cold enough to make it worthwhile.

Although the combustion products of all these coal-burning fireplaces severely contaminated the air of cities, the flues provided generous indoor ventilation, especially after the sanitation- and fresh-air-conscious Victorians introduced a requirement for a direct air entry into each habitable room. Simple gratings in the form of air bricks in the typically solid 9- or 13.5-in. brick walls became universal and these and the fireplace flues, together with general leakage, could provide a dozen or more air changes per hour. They astonished foreigners, but condensation seldom occurred.

In the 1920s and 1930s the usage of fireplaces diminished as skivvies disappeared but, as we said, in the multi-flue chimneys it needed only one warmed flue to power them all, and no amount of water vapour produced indoors could catch up with this rate of removal. Chilliness was commonplace but was slightly eased by the introduction of the familiar 11-in. cavity wall, affording not only better protection against damp penetration but modestly better insulation, as well as the lower thermal capacity of the single inner leaf.

Changes were also taking place below ground. The traditional basements of urban housing lost their economic justification as suburbia developed and the customary ground floor became boarding, usually softwood, over a ventilated void but the quality of softwood gradually deteriorated from the 1920s into the 1930s as commercial growth replaced natural in the exporting countries. As a result shrinkage began to provide additional ventilation through cracks between boards, amounting sometimes to as much as two or three square feet of opening – 0.15–0.25 m² – in a large room. This particularly applied to low-cost housing, and the resulting air changes added materially to the growing discomfort of its occupants.

After 1945 there was a drive to increase thermal comfort and reduce the wastage of heat up chimneys. Fireplaces and their flues commonly dropped to one per dwelling (after much social anguish, because the usual four were widely perceived as a kind of status symbol), and the fire itself was placed in one of several kinds of slow-burning stove or grate. The air brick requirement in the walls of habitable rooms was abandoned as a peculiar hangover from Victorian times, and because of post-war economic pressures to reduce timber imports the suspended floor was replaced by concrete directly on the ground. These changes combined to reduce ventilation rates dramatically and certainly produced greater thermal comfort, but they were accompanied by the first complaints of condensation. Permanent ventilation made an uncertain return in the form of closeable slots in window frames but these were often sealed off by householders unable to understand the connection between ventilation and condensation and more concerned to avoid draughts and the apparent waste of heat (Figure 2.3).

In the late 1950s there was a further departure from earlier conventions when political and economic pressures led to demands for higher dwelling densities in urban areas in the form of medium- and high-rise

Figure 2.3
Condensation in houses almost hermetically sealed by aluminium roofs and wall cladding.

flats, and innovative prefabrication using concrete cladding was found to help make this economic. Heating without fireplaces and flues became a practical necessity, and inexpensive fan-blown ducted warm air systems were introduced, usually getting their heat from individual heat exchangers on a central circulating supply of hot water. Similar systems were soon also widely adopted for houses. They had the superficially attractive characteristic that they could be controlled by clock thermostat, delivering heat when people were at home but not when they were out, apparently thereby offering the economy of tailored heating.

This was the late 1960s, and in 1967 disaster struck. There had been no requirement for fresh air to be drawn into these ducted systems, there were no flues to allow the escape of the accumulating water vapour or to generate ventilation, and concrete cladding restricted migration through the outer walls. Vapour escape through party floors and walls was frustrated by equal vapour pressures in the adjoining dwellings.

The RH values cycled up and down at higher and higher levels as temperatures rose and fell with the coming and goings of occupants, gradually depositing more and more moisture in the colder intervals until no more could be held. Wall, ceiling and floor surfaces became damp or wet, wallpaper was stained and even fell off, curtains, carpets, clothing and bedding became damp, wood furniture warped and split and electrical faults occurred. Condensation became headline news for three winter seasons as the people concerned endured misery and the contents of their homes suffered extensive damage.

With hindsight, one can see easily enough what had not been apparent during the years leading up to this tragedy. Each change to provide more heat and to waste less of it had seemed beneficial in itself but had been viewed in isolation, and what each change was also causing was a step-by-step loss of means of escape for water vapour until, as remarked earlier, it was often being created by family life at rates much exceeding any possible rate of escape. Most unfortunately the amount of moisture generated by family life was not known in Britain until 1969[6] and therefore no necessary rate of escape or removal had been recognized. The

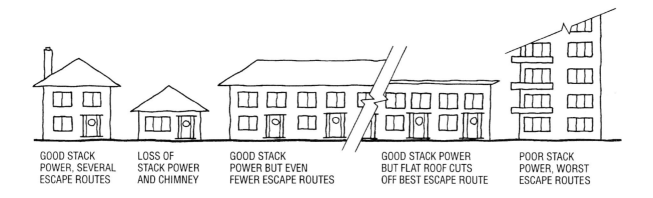

GOOD STACK POWER, SEVERAL ESCAPE ROUTES   LOSS OF STACK POWER AND CHIMNEY   GOOD STACK POWER BUT EVEN FEWER ESCAPE ROUTES   GOOD STACK POWER BUT FLAT ROOF CUTS OFF BEST ESCAPE ROUTE   POOR STACK POWER, WORST ESCAPE ROUTES

daily vapour production figure then established of 7–11 litres per day per family is now familiar.

Figure 2.4
Housing types in rank order of risk from condensation, reading from left to right.

A rank ordering of condensation risk

With this brief history in mind, some simple design facts about different dwelling types enable us to set them in a rough but informative order of condensation risk (Figure 2.4).

At the lowest level is the two- or three-storey free-standing house with vented cavity masonry walls, vented tiled or slated pitched roof, an open fireplace and no vapour barriers. The envelope is wholly exposed, the walls are moderately permeable and the roof more generously so, stack power overall is strong because of the two or three storeys, and there is good flue ventilation.

Next comes the single-storey house of similar construction, differing only in having less overall stack power to generate ventilation.

Then in third place there is the terrace house. The benefit of two of the external walls has been lost because the vapour pressures on the other side of party walls prevents loss through them, and if the front and rear walls are generously glazed, the glass will be another barrier.

In fourth position are houses where felted or asphalted flat roofs have replaced tiling. This cuts off one of the better escape routes for vapour, and if wall insulation requires a vapour barrier, loss through walls also gets reduced. The omission of a fireplace and its flue will remove the best escape route of all.

Lastly, at highest risk are flats, enclosed by party floors as well as party walls, and if the exposed area of the envelope is again generously glazed and/or the cladding is of concrete or metal, the opportunities for natural escape of water vapour almost disappear. There will seldom be a fireplace, but there will usually be powered air extraction from kitchens and internal bathrooms, and if this operates continuously it will remove moisture effectively, but if it operates intermittently, on a light switch for example, its value is much reduced.

Cooking by gas increases condensation risks because of the water vapour produced by its combustion, but modern kitchen design usually ensures its quick discharge to atmosphere by extract fans.

In terms of geographical location, the higher the Driving Rain Index, the greater the condensation risks will be because of the high environmental humidities.

Sadly, the worst humidity conditions are often linked to poverty, with high-density occupancy of old property, little heat input, no avoidable ventilation, blocked fireplaces, not much insulation, unvented cooking by gas, and little understanding of why all these combine to produce such bad conditions. Illness and despair are natural consequences.

Removing moisture by ventilation

As opportunities for vapour escape through the envelope diminish, dependence upon its removal by air changes increases. To make design judgements in this respect we need information about amounts required and a set of assumptions to establish a reference base, i.e.:

1   That a family of four may generate as much as 11 kg of water vapour per day and that the aim will be to maintain an air temperature of about 20°C at an RH of about 50–60 per cent.
2   That the least efficient air for passive ventilation occurs in winter when low air temperatures outdoors of the order of 2–4°C commonly occur, and outdoor RH values of at least 85 per cent will prevail for longish periods.

Such air is so wet that the least volume of it which must then be changed to effect the necessary removal of indoor vapour will be of the order of 75 m³ per hour, which means about one-third of an air change per hour for a dwelling of around 90 m² area and one-seventh for a dwelling of 200 m². However, the larger the dwelling, the greater the envelope area will be and the greater the loss by migration and leakage. Dew point on indoor surfaces would occur at about 11°C under the assumptions just described.

Although the assumptions are not unreasonable, in many dwellings 20°C will not be an affordable norm in winter or may not be affordable in all rooms. Also, moisture production rates will sometimes exceed 11 kg per day with occupancy by more than four people, and external RH values will often be above 85 per cent in winter, close to 100 per cent at around 0°C. The ventilation air will then be even less efficient, the vapour production rate high and temperatures undesirably low, and the condensation risk will rise to a near-certainty. Bedrooms are often at a particular risk because two people asleep for 8 or 9 hours in a poorly heated room on a cold night with door and windows shut will produce very high RH values and a large amount of condensation by morning.

Case note

One such case concerned a bedroom occupied by two teenage boys. The walls were of 9-in. brick to which the local authority had added 1-in. fibreboard on battens for insulation on the inside, finished with a skim coat of plaster. The room was unheated and the boys kept the window and door shut for warmth. Vapour easily passed through the plaster and fibreboard lining and condensed in large quantities on the original brick wall, now colder because of the inside insulation, and condensate flowed in large amounts out on to the floor and saturated the carpet.

Using stack power

Stack power is the passive energy source always ready to effect air changes when inlets and outlets are available to let it operate. Inlets and outlets should be roughly equal in total area, and experience suggests that for a family of four in a small house this will be in the region of 250–350 cm².

Inlets can be the most problematic because of draughts. A modern ducted warm air system will be designed to bring fresh air into the mixing chamber on a balanced basis, and this is essential. If radiators are the heat source, small inlets are occasionally positioned behind them so that they warm the incoming air. Fan-powered sound-insulating units are available which can warm the air and are used for dwellings near airports and other major sources of outdoor noise. A desirable development might be a small simple inlet with electrical tempering. There are, of course, windows with air inlets in their frames, but these are too easily blocked off. Doubtless 'intelligent' houses will in future offer controlled warming of inlet air. Meanwhile, leakage and occasionally opened windows and doors usually provide the needed air entry without other provision.

Bedrooms are strong candidates for inlets because of the high condensation risk that arises when windows are kept shut because of noise or, on ground floors especially, the risk of vandalism and burglary, and doors are often kept shut for privacy.

Inlets in the envelope should be positioned as low as practicable and outlets placed high so as to maximize the height of a stack. Where there is a pitched roof it is now an increasing practice to insulate the loft floor and ventilate the loft itself, and it then makes sense in bedrooms to release the stack power by openings through the ceilings. If there are two or three storeys and if more ventilation is judged necessary, the upper part of a stair hall is another good outlet position. Water vapour passes slowly but steadily through most ceilings anyway, and venting is simply a means of accelerating the escape.

Loft ventilation is discussed with pitched roofs in Chapter 10.

Case note

In the case of my own house, built in 1947 in a dryish part of Britain, I insulated the ceilings with 100 mm rockwool and vented the bedrooms and bathrooms direct to the roof space. The roofs are tiled over conventional battens and sarking felt. There is some inadvertent eaves ventilation due to wood shrinkage and two gable ends are vented. The house has central heating and one fireplace and typical winter RH values are 35–45 per cent. The roof timbering stays dry and there has been no evidence of dampness anywhere.

Ducts as flues

Reference has been made to venting by means of ducts as a flue system, and it is sometimes recommended that ducts should drain upwards directly from kitchens and bathrooms through the roof space to atmosphere.[8] To my mind, this is not logical. Modern kitchen design usually provides for direct extraction from cookers to the open air, and the production of vapour in bathrooms is only occasionally intensive. Neither bathrooms nor kitchens function as good gathering points for water vapour because the lifestyle of many people requires their doors to be kept shut. If ducting is being considered as a ventilation technique, it seems better to duct from a central location like a stair hall to which water vapour from other rooms always gets some access, or direct from bedrooms, as previously suggested.

A flat roof has the evident disadvantage that it lacks a roof space to exploit in these ways. It may then be best to vent a central position direct to atmosphere; but see also the discussion in Chapter 9.

Communal residences

College residences, hostels, barracks and other communal residences usually have the highest density of occupancy and its most prolonged use. They typically comprise large numbers of bed-sitting rooms, with communal baths and showers and often communal cooking facilities. There is probably the equivalent of full occupation of such buildings for an average 12 hours or more per day, and consequently high moisture production.

It is necessary to arrange for the removal of this humid air. One possible way is by providing individual room ducts operating by stack power but a more positive method is by continuous ducted extraction by power at roof level, where the plant can be accessible for maintenance. Communal washing and cooking facilities need a similar but separate arrangement to avoid any risk of short-circuits.

Where there is dependence on fan power, it is essential to provide adequate monitoring and maintenance of motors and regular replacement of fan belts, with some sort of warning system when failure occurs. Severe misfortunes, including major roofing failures, have occurred when monitoring failed and fans ceased to function properly for several weeks or months, leading to wood rot in the framing and decking.

Case note

Some four-storey university residences were designed with the rooms in a square around common baths, showers, toilets and kitchenettes, with extract fans on the roofs. These had deep timber frames carrying asphalt on chipboard. Venting of the roof voids was by louvered openings on all four sides of the building, but air and vapour movement was evidently too restricted by the roof framing to take place effectively. When fan belt failures went unnoticed for a period the chipboard in the area over the service rooms deteriorated and collapsed (Figure 2.5).

Figure 2.5

Flats

Flats generate the highest humidity risks. Individually they not only have little stack power to help natural ventilation but no easy way of using it. They have minimal external wall area and in high blocks they need good standards of air-tightness in windows and balcony doors to cope with their greater exposure to wind.

Take low-rise blocks first. If they have internal bathrooms and/or kitchens, these will be expected to have powered air extracts of their own to deal with odours and vapour concentrations. These will usually be ducted to roof outlets and the fans will probably be triggered by a light switch, perhaps with a timed overrun.

The questions that then need thought are the sources of air to balance the extraction and how well this will remove air and vapour from the bedrooms and living rooms. If ducted warm air heating is used and fresh air in proper quantities is drawn into the mixing box, the necessary air changes can take place with the kitchen and bathroom ducts completing the circuit; but if other heating is used and/or if no other provision is made for air entry, the other rooms may not keep dry. The problem deserves some proper systems thinking in design.

With medium- and high-rise flats one moves into the field of mechanical services design where the common practice is to use a continuously operating air extract system with ducting to fans on the roof. The continuous operation does a better job of general ventilation of other rooms besides the bathroom and kitchen, but this can be made positive by supplementary ducts from the other rooms.

2.13.2 Rooms for occasional meetings

Places where community meetings, local committees and various informal gatherings are held typically get intensive usage but only for a few hours. At the average meeting there will be a large short-term injection of human breath, perhaps some damp clothing, and probably some steam when coffee is made. Most of the buildings used for such venues are built with walls of high thermal capacity and often have only temporary heating. If meetings are prolonged without adequate ventilation in cold weather, conditions can become thoroughly unpleasant with condensate forming on ceilings, walls and windows. The room will feel, and be, unhealthy. Several ailments are carried on human breath and transmit more readily in damp than dry air.

A better design basis needs to be found for this type of accommodation. Good sense points to the use of a low thermal capacity envelope, with adequate insulation so that it can warm up quickly when intermittently heated, together with some simple arrangement for ventilation.

2.13.3 Office buildings

For smallish office blocks where air conditioning may not be economic, architects have tried various treatments but few are successful. It may be acceptable to assume that no powered ventilation is required where the offices are individual rooms, because sensible window design can provide for personal needs and tastes, though external noise must then be acceptably low. Larger office blocks, perhaps with open planning as an option, are another matter: they must normally be air conditioned, and any assumption that this may not be necessary can have dire results.

Case note

The worst case of office condensation I have seen was in a medium high-rise, mainly open-plan block very exposed to west coast weather. The building, in which about 500 people were working, was heated by radiators. There was no powered ventilation, and in cold wet weather no windows were ever willingly opened. The building had an exposed concrete frame with no thermal breaks. Condensation damage to plaster around windows and on the wall, floor and ceiling finishes generally was extensive, and working conditions were almost unbearably damp (Figure 2.6).

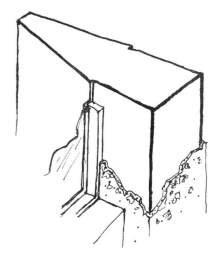

Figure 2.6

The design of air-conditioning systems is a matter for services engineers, but investigations of roofing problems on a large block of high-quality offices brought to light a point about air conditioning that is of wider relevance in building design.

Air conditioning normally includes humidity control which can keep RH values below, say, 55 per cent, which means in effect that it will remove excess water vapour from the building. What was found on this roof was that fibreglass roof board which had been sandwiched between the vapour barrier and an asphalt weather finish gave the latter poor support so that it suffered fractures. The fibreglass became wet by leakage but local areas of dry insulation and concrete were found where parts of the vapour barrier were missing. The only rational explanation seemed to be that the air conditioning had been able to do this because the missing barrier allowed the excess dampness to migrate downward through the concrete to be vented away.

It is sometimes argued that in climate zones where mid-January air temperatures typically average out above 1 or 2°C, there is no need for a vapour barrier in buildings where RH values will be kept below 55 per cent or 60 per cent. See also Section 2.4.

2.13.4   Industrial buildings

Many different humidity situations arise in factories. The occupation density may be low or high and the processes may be dry or produce moisture which may be local or general and aggressive or non-aggressive, and there may also be requirements for specific temperatures and humidities. A few factories are built specifically for one type of production and are written off when it is finished, but most are built speculatively and need to be adaptable. No factory building is likely to be built now with a clear height of less than 5 m, but it will be more adaptable for modern industry if the height is 6.5–7.0 m. Warehousing may require 25 m or more, resulting in large air volumes.

Here are some examples of moisture-making industries.

Low aggression
Breweries and distilleries
Many kinds of food production
Laundries
Concrete pre-casting
Paper manufacture
Natural and man-made fibre manufacture
More aggressive
Pickling and plating of metals
Chemical and pharmaceutical production

Paint and finishing shops
Activities needing processed air environments
High-quality colour printing
Some types of leather processing

Buildings are much the cheapest item in the overheads of manufacture, and reasonable adaptability costs little and protects their market and manufacturing value. Factory users generally must be able to accommodate whatever changes of plant and processes are needed to match competition in terms of lower cost or higher quality production, or risk going out of business.[9]

Adaptability in this context is mainly a matter of roof design to get adequate height and structural load-carrying capacity. Then, to control humidity, a roof type is needed which can be perforated readily for air-processing equipment and then be made reliably watertight again. It must be able to carry air conditioning or other air-processing packages with their associated weight and ducting, all of which indicates the need for flat or very low pitched roofs.

Factories often have to stand empty for longish periods awaiting a sale or lease. They are presumed to remain dry, but this is unlikely and a case illustrates the reasons.

Case note

Some speculative factories were built with low-pitched profiled aluminium roofs in a wet climate zone. Plasterboard ceilings carried some fibreglass insulation. The factories had proved difficult to let, but one was occupied and had a high population density and only fortuitous ventilation. RH values were high and plasterboard was sagging. Both condensate and leaking water dripped into the production area.

Other units had stood empty and closed for several years, and when a letting was obtained for one of them, it was found that the plasterboard had dampened and sagged and some of it had fallen out.

What caused this was that in winter the external air temperature would sometimes get down to freezing or below, with RH values of 80–95 per cent, and RH conditions indoors would try to become similar. Insulated roofs of low thermal mass will often be even colder than the inside air due to radiation loss to cold skies, so that RH values of the air close beneath the roofing will rise to near-saturation. In this cold, damp air a material such as plasterboard will absorb water vapour and gradually soften and weaken, sagging bit by bit until perhaps it falls out of its supports.

Factories often stand empty for longish periods, so it is wise to avoid the use of damp-sensitive materials as linings.

2.13.5 Indoor swimming pools

Indoor pools usually contain chlorinated warm or hot water and evaporation and vapour formation take place. The rate at which this happens will depend on the surface area of the water and its temperature[1]; it can be quite rapid. High RH values will become prevalent in the parts of the building accessible to pool air and the vapour will contain some of the chlorine. This is potentially corrosive to metals and sometimes reacts adversely with other building materials. It is important to check with the manufacturers whether particular materials proposed for use in the building fabric will be at risk in this way.

In circumstances where so much vapour is going to be produced, careful thought needs to be given to the design of the building envelope. Outward migration through it should be encouraged but care must be taken to avoid interstitial condensation in it. It will often be sensible to assist or replace natural ventilation by powered air changes to keep RH values at reasonable levels and engineering advice should always be obtained. The behavioural characteristics of particular types of walls and roofs are discussed throughout this book.

Where a pool building is air conditioned, the pool system must never be linked with the system for public or other rooms. The smell of chlorine is pervasive and can be very objectionable, as one or two hotels have found to their cost.

2.13.6   Museums and art galleries

For the managers of museums and art galleries the prime custodial duty is conservation, and for organic displays and sometimes inorganics such as ancient glass or clayware, the quality of the indoor environment is crucial.[10] Apart from lighting, its most important requirement is constant or only gradually changing RH and temperature. In a temperate oceanic climate it is usual to aim at about 55 per cent RH plus or minus 3–5 per cent and an air temperature of about 18°C, not because of any specific merit for conservation but because they are the values that can be maintained with least energy input in that kind of climate.

In eastern areas of North America RH levels as low as 40–45 per cent are regarded as desirable museum norms. Running costs of plant can be significantly affected by too large a gap between indoor RH values and the outdoor norms, and there is a case for using sophisticated controls that allow very gradual changes to take place indoors as average seasonal values change outdoors. It is abrupt changes which cause the greatest damage to artifacts. The longevity of sensitive artifacts in favourable burial conditions is well known, and so was the excellence of their storage in caves in Britain during the Second World war.

Achieving these standards is a test of skill for building services engineers, who have to design not only for these broader exigencies of climate but also for the often abrupt changes of temperature and humidity caused by the arrival and departure of the public in quantity.

Sometimes the breakdown of equipment or power loss can cause significant and even irreparable damage to artifacts and there is much to be said for introducing stability 'flywheels' into the design of the building fabric, both indoors and in the envelope, to do some tempering. For example, moisture-absorbent materials can act as reservoirs to take up or release vapour at a moderate rate, while a building envelope with high thermal mass, protected externally by insulation, can help to achieve constancy of temperature. (See also the discussion under 'rainscreen systems' in Section 4.8.)

Case note

In one well-known American gallery I was told that the main building, constructed in the 1930s with thick marble-faced walls and limited roof glazing, could hold a stable indoor climate with only about 3 hours of air conditioning per day, while a 1970s addition, with a far lower thermal mass and large glass areas, required about 21 hours' input to achieve a similar stability.

At first sight, it may not seem to matter much whether it takes a little or a lot of conditioned air to maintain a stable climate indoors, provided that this can be achieved and afforded. However, the more elaborate and extensive the system, the more vulnerable the collection will be to poor plant maintenance, occasional shut-downs, and the implementation of ill-advised cost-cutting exercises.

Matters of conservation and design policy such as these will usually require the agreement of museum or gallery trustees and managements and need to be established in advance because of their implications for architectural design and engineering services. This underlines the need for a particularly good rapport between architect and engineers while the design concept is being formulated and developed.

Because of the importance of controlling the indoor climate for conservation, thorough commissioning of the plant must be carried out before it is accepted. Commissioning should be done in two stages: the first when the installation and building have reached practical completion, and the second when there has been a full season of operation and monitoring records are available for study.

Case note

In the operation of a new gallery installation it was found that specified standards of delivered air were not being met. Pressure loss was taking place by leakage from builders' ductwork. The question naturally arose whether the fault lay at the door of the builder or the architect or the consulting engineer, emphasizing the importance of close collaboration between architect and engineer during design, specification and testing.

2.13.7   Historic buildings and collections

In respect of moisture management, historic buildings are special in that both the contents and the fabric usually require conservation. The historic buildings of Britain and Ireland have generally survived quite well, due probably to having walls with a high thermal mass, good ventilation by generous fireplaces, and relatively small numbers of occupants in relation to large enclosed air volumes. The conditions were not ideally constant but they changed only slowly. On the debit side, however, there was no damp-proofing of floors or walls at ground level, and the walls often had rubble cores that acted as wicks, pulling moisture up from the ground which by transpiration was able to enter the indoor air as well as the outer atmosphere. Since the 1950s, many of these historic buildings have also received large and abrupt moisture and thermal loads due to the influx of visitors at specific times, and risks to conservation have risen correspondingly.

The small ancient churches that grace the countryside of Britain have suffered another change in modern times that has often led to trouble. Traditionally they were ventilated by leaving the entry door open, with screen doors to exclude birds. They had no heating and the heaters that began to be installed in the nineteenth and early twentieth centuries were operated only when a congregation was to be present, not to provide constant warmth. Ventilation then began to be seen as something that wasted heat and the open doors also exposed the church to increasing vandalism. The result was that the combined warmth from the congregation and the heaters in winter raised air temperatures and water vapour levels beneath the roof, where the timbers and metal or tiling would be cold. RH values around them would rise rapidly and cause condensate to form. Corrosion risks would rise (see Chapter 10 for the main discussion of corrosion) and the moisture content of the timbers would gradually increase towards a level favourable for rot and for the propagation of death-watch and other beetles, with a subsequent need for repairs or replacement. The moisture content of timber must be kept below 20 per cent for safety against rot and beetle attack, as remarked earlier.

Heaters fuelled by gas or oil were often flueless and this introduced further considerable quantities of water vapour. Coal-burning appliances released pollutants which could contaminate the humid air and sometimes initiate the corrosion of roof metal. The problems created by these heaters and the alternatives now available have been discussed by Bordass.[11]

The great historic mansions gained much of their ventilation and vapour escape by virtue of their large fireplaces and flues and by their large envelopes where much air and vapour leakage occurred. Although the flues usually had draught dampers, the fireplaces were rarely blocked off completely so that some air and vapour could usually depart by this path. This is probably what is coping with the moisture loads of visitors today. Conditions are now usually well monitored, but the moisture loads are large (of the order of 8000–10 000 litres per 100 000 visitor hours), and unless avoiding action is taken, this may create trouble in the medium or long term.

2.13.8   Key points for designers

- Buildings that do not have air conditioning must be able to lose water vapour by exfiltration, outward migration, or by direct escape routes.
- Air-conditioned buildings generally need to avoid all these escape mechanisms and rely on the removal of excess vapour by powered air changes.
- In dwelling types, the lowest risks of condensation are found in free-standing dwellings with pitched roofs, tiled or slated, with an open fireplace and flue. The greatest risks lie with multi-storey flats with no loss through party walls or floors, low-permeability cladding and no fireplace or flue.
- The most exacting types of occupancy are art galleries, historic places, and museums; These need very stable temperatures and RH values for conservation but are assailed by the abrupt arrival and departure of heat and vapour loads in the form of large numbers of people. They need well-sealed buildings, preferably with a high thermal mass envelope to assist thermal stability, and carefully designed air-conditioning plant which has been commissioned properly. Specialized laboratory and storage buildings may need even more stable indoor climates but do not have large influxes of visitors.
- All other accommodation lies somewhere between these various extremes and designers should give thought to just where a given project lies within this spectrum.

- The key to clear thinking about water vapour is always to keep in mind that it is a gas exerting pressure, and higher pressures will always try to flow to lower by migration or leakage in efforts to reach equilibrium.

References

1   Milbank, N.D., 'Energy consumption in swimming pool halls', BRE Current Paper 40/75. His formula for emission from a wet surface is: $16 \times$ area (m^2) $\times$ SVP at water temperature – VP of surrounding area $= x\ q/h$.
2   BRE Digest 210, Principles of Natural Ventilation, February 1978.
3   Brundrett, G. W. and Blundell, C. J., 'An advanced dehumidifier for Britain'; *The Heating and Ventilation Engineer*, November 1980, pp. 6–9.
4   BRE Digest 370, Control of Lichens, Moulds and Similar Growths, March 1992.
5   BRE Digest 299, Dry Rot: Its Recognition and Control, July 1985.
6   Allen, W. and Rich, P., 'Condensation: a warning to architects', *Architects Journal*, 14 October 1970, pp. 907–9.
7   Brand, R. G., 'High humidity buildings in cold climates: a case history', Paper 934 of the Division of Building Research of the National Research Council of Canada 1980, price $1.00.
8   Gaze, A. I., 'Passive ventilation: a method of controllable natural ventilation of housing', Report No 12/86, UK Timber Research & Development.
9   Allen, W. A., 'Factory design for the future', *J. Inst of Production Engineers*, June 1960.
10 Thomson, G., *The Museum Environment*, 2nd edn, Butterworth, 1986.
11 Bordass, W., *Heating in Your Church*, Church Information Office, 1984.

Further reading

1   Lacy, R., 'The climate inside King's College Chapel, Cambridge', *Studies in Conservation*, **15**, No. 2, 1969, 65–80. A most interesting account of air behaviour in a high stone chapel with much air leakage through large windows.
2   BRE Digest 210, Principles of Natural Ventilation, 1978.
3   Uglow, C.E., Background Ventilation of Dwellings; A Review, BRE Report, 1989.
4   BRE Digest 336, Swimming Pool Roofs; Minimizing the Risk of Condensation, 1988.
5   The Chartered Institute of Building Services (CIBSE) has a guide, published in several volumes or in individual sections covering heating and ventilation technology in detail for engineers.
6   Dubin, F.S., 'Intelligent building', *Construction Specifier*, February 1990. Deals, *inter alia*, with RH levels for electronic equipment, and also indoor air pollution by office equipment.

Part Two
The envelope

3 Ground water, basements and ground-level works

The principal problems for the envelope below ground or resting on it relate to water or moisture in the subsoil, though an additional concern in some parts of Britain is radon gas. The discussion is taken in three parts: the first on subsoil water, moisture and radon, the second on basements, and the third on ground-level construction.

3.1 Subsoil water and radon

3.1.1 Subsoil water

Ground water is a stage in the hydrological cycle, which begins with rain and ends with stream flow run-off or storage in deep aquifers. There is water lurking at some depth below most ground, especially in an oceanic climate. If it is close enough to the ground surface it can cause problems during construction and subsequently affect foundations and basements.

The water is located between soil particles and flows through the interstices of the soil as a result of gravity or pressure differentials. The flow can be rapid in open-textured soils like sand and gravel, but the finer the texture, the slower the movement. In clay it can be very slow. Conversely, sands and gravels usually have high shear strengths and little compaction so that they carry loads well, while water in clay behaves somewhat like a lubricant, allowing the soil to be displaced under excessive loads or to shrink if drying takes place.

When evaluating a site for wetness, a check of the following points will allow a useful first opinion to be formed.

Local road names

Names such as Mill Way, Pond Lane, Well Walk may be significant and should alert a designer to risks of wetness.

Drainage ditches and pond sites

If ditches and ponds are dry when inspected, find out whether they re-wet seasonally or in storms.

Local water table levels

Check the seasonal variation and trend of movement of water table levels locally. In the past these have tended to fall in Britain, but recently they began to rise in some areas because industry now uses less water.

Underground streams

Check for underground streams. Some may be running in conduits, but streams never disappear and conduits rarely contain all their water. Building over conduits is seldom allowed. Also, if there is a rock base below ground, check for water-bearing fissures.

Pollutants

Some soils contain pollutants which can attack concrete or metals or be dangerous to health. Check with the local authority or have tests done. Sulphates are a particular risk, notably of calcium, magnesium and sodium, and if soil water contaminated by them stays in contact with concrete, it can cause cracking and expansion.[1] Landfill sites in particular are likely to be contaminated and may also be poorly consolidated. If so, they may settle or compact unevenly under load and cause pipes or drains to fracture. Broken water mains can cause serious misbehaviour of the subsoil and foundations and may spread contaminants.

Trees

Trees de-water soils by transpiration through their foliage. Young trees are particularly thirsty, although they only cause drying over limited areas. Mature trees take up moisture more slowly, but affect larger areas over longer periods.

Trees with a globular shape usually have globe-shaped root systems, whereas tall, narrow trees develop wide root systems to resist overturning forces in wind (Figure 3.1). When mature trees are removed from a clay site this drying system goes with them, and the ground gradually re-wets and expands both upwards and sideways, sometimes taking eight or ten years. It can lift two- or three-storey buildings and bend or break them, although there is some engineering technology by which it can be mitigated. It can be disastrous, especially with lengthy buildings, for they are not strong enough to resist being bent on unevenly expanding clay or refill sites. I have seen upward bends of 50 or 75 mm.

Sloping sites

On sloping ground, water is generally moving down the slope below the ground surface. The flow may be diminished by barrier drains, but these are not always effective for very long. It is unwise to assume that the flow can be stopped in this way above a site on a slope. There may also be a risk that absorbent ground may be replaced by a non-absorbent surface such as concrete, causing destabilizing run-off. Check where it might go.

Foundations, basements and retaining walls on sloping sites will act as dams and impound some of the water. As a result, undesirably high hydrostatic pressure can develop which may break into basements unless effective drainage is provided. The load of a building can compress soil sufficiently to cause partial damming.

Think hard about the placing of soakaways. If tree roots can get into them they will find the water in the drains enticing and may seek out its source, fattening and filling the drains in the process.

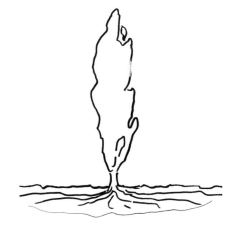

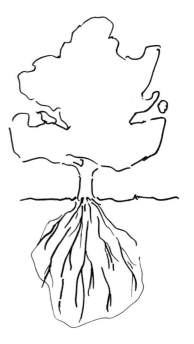

Figure 3.1

When evaluating a site, refer also to the appropriate large-scale Ordnance Survey map, dig trial holes or take bores, talk to the local authority and, if appropriate, to statutory suppliers of public utilities. If the design of a basement requires structural engineering advice, the architect and consulting engineer should evaluate the site together. It may be necessary to compile data throughout a season, and the cost of this should be borne in mind.

The following are three cases of insufficient forethought about water on sites.

Case notes

A hospital was built on a clay site which had a long history of ponding from springs. The site was not adequately researched. A tall ward block was put on deep piles, but a single-storey-plus-basement block of offices and laboratories connected to it along two sides was founded only on pads. The movement joint system along the two interfaces could not accommodate the restlessness of the two-storey construction in relation to the immobile ward block. Leakage and internal damage became chronic and impossible to prevent (Figure 3.2).

A second case concerns a very high building built on London clay. This also had a two-storey attachment, but this was below ground and had a public garden over it. London clay is very deep, and under a high building it is usual to use a deep basement excavation so that the weight of the removed clay contributes to reduce the subsequent shear stress in the lower supporting clay when the load of the finished building has developed on it. This sets in train a particular sequence of movements. The removed weight of the excavated clay allows the lower clay to decompress and rise until the load of the building gets placed on it, causing gradual recompression and downward movement. The latter process takes several years and the movement may be 50 or 60 mm each way on London clay. Other clays behave similarly but amounts of movement will vary.

In this case the attached two-storey underground block had insufficient weight to cause much recompression of the clay beneath it and, unfortunately, it had been securely connected to the slowly moving high block. When this eventually began its downward movement, it bent and cracked the two-storey part, causing numerous leaks. The problem was not understood until after the high block had come to rest. Repairs were very expensive.

Figure 3.2
A perpetually restless junction.

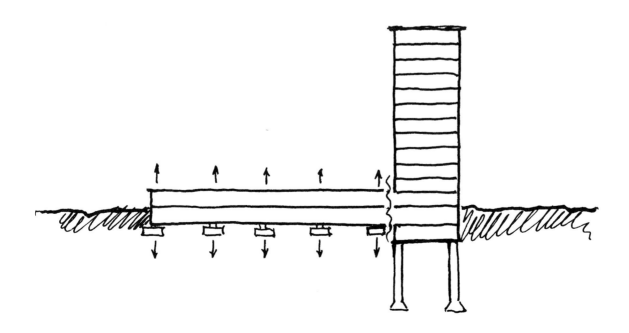

In the third case, a group of connected low-rise buildings was built on a slope. Field drains feeding the slope above the site were supposed to have been stopped by a barrier drain. All the buildings rested partly on retaining walls and partly on strip footings, mostly crossing the slope. The building at the bottom of the slope was a long, high hall, also set transversely. The side away from the slope faced south, where there was a growth of young trees (Figure 3.3). Water seeping down the slope was impounded in large amounts by footings and retaining walls. It leaked through the stairs in the buildings, forming ugly accretions of dirty calcium carbonate. After several dry seasons there was an unexpectedly wet autumn, and the south wall and one corner of the long hall at the bottom suddenly began to move towards the trees, threatening a collapse of the entire structure. This was averted by propping inside and out, done at some danger, after which the south wall was given new supports by needles cantilevered from trios of external piles, and structural remedies were carried out.

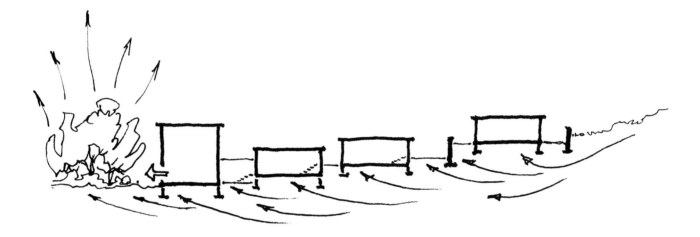

3.1.2 Radon protection

Figure 3.3

Radon is a carcinogenic gas produced by the radioactive breakdown of uranium and is present to a small extent in all soils and rocks. It mixes with air and water in the soil and seeps out of the ground as a gas, usually diluting harmlessly in the open air. However, where amounts are larger and buildings are present over the source there is a risk of seepage into them, and this can result in small radioactive particles being inhaled, which can create a risk of lung cancer. Fortunately the technology needed for protection against radon entry is similar in principle to that needed for moisture protection, which is discussed below.

In Britain, the highest-risk areas for radon are Devon and Cornwall and some parts of Somerset. Lower but significant levels occur in most of Wales, Gloucestershire, Northamptonshire, Derbyshire, south-west Scotland and the north of Scotland. Local authorities should have local data on risk levels.[2]

3.2 Basement design

3.2.1 Movement joints

Structural movement joints in basement envelopes are high-risk positions on any wet or damp site and should generally be avoided. It is not necessary to use them to accommodate differential thermal movement because at basement depths soil temperatures are stable, but they are necessary where differential settlement has to be accommodated between parts of a building that have been founded differently. The shrinkage that naturally occurs as concrete dries should be restrained and distributed by reinforcement and not concentrated and accommodated by the use of movement joints.

3.2.2 Options for basement design

There are four alternative approaches to basement protection against water, vapour and radon penetration:

- Use of good-quality concrete without tanking but usually with drainage around the footings.
- Use of a tanking membrane applied externally.
- Use of a tanking membrane applied internally, as part of a sandwich.
- Use of a cavity in floor and walls, with cavity drainage.

The second and third options, and sometimes the fourth, can ensure a dry basement even in wet soils, but the acceptability of the first and cheapest option, an untanked basement, is essentially a matter for judgement.

3.2.3 Basements without tanking

Before concrete was widely used, basements in Britain had walls of brick or stone masonry with earth or paved floors. Such dryness as they enjoyed was due to a combination of dry sites and free indoor evaporation from the walls and floor. However, they were often very damp, and their evaporative cooling made them excellent pub cellars or wine stores.

Today, concrete has become the usual material for basement walls and floors in Britain and elsewhere. Whether unwanted moisture will get into an untanked basement depends then on several factors. Free water in the soil must stay well below the slab, the concrete must have low permeability, there must be as few daywork joints as possible, and uneven structural settlement that would open these joints must be avoided. Reinforcement is an obviously useful insurance against this risk.

Radon can diffuse through concrete, but only to a limited extent and at present no cases are known where it has led to high radon levels.

Concrete quality is discussed mainly in Chapter 4, but a few points can usefully be made here:

- The aggregate must be well graded, of good quality, and free from porous stone.
- Placing and consolidation of the concrete must be done properly.
- Initial curing of the concrete must be controlled to avoid shrinkage cracking.
- Reinforcement should be used to minimize shrinkage of the concrete and it will also serve to enhance the integrity of the basement.

- To minimize shrinkage, reinforcement should be placed near enough to the surface to counteract the effect of early drying while being nevertheless adequately covered. Weight for weight of metal, smaller steel in a closer mesh is more effective in controlling surface shrinkage than heavier steel in a more widely spaced mesh.
- Reinforcement should be carried across daywork joints for these might otherwise open due to shrinkage or structural movement.

Basement concrete should be specified with the aim of minimizing shrinkage and permeability and the following should be noted:

- Ideally the proportions of aggregates and sand should be such that there are only minimal voids to be filled by the cement, and
- The cement content should then be kept as low as is consistent with the requirement of strength and resistance to attack by soil-borne sulphates.
- The water content should be kept as low as proper compaction requires, because permeability tends to increase dramatically as the water content of the mix increases, and permeability increases vulnerability to sulphate attack and to infiltration by water vapour and other gases. The infiltration is slow, but it can be accelerated by upward pressure by water or vapour under buildings. In the case of radon, the rate of infiltration can sometimes reach several cubic metres per hour if underpressure exists in a building, for this helps to draw in soil gases, but this is an unlikely situation in domestic property.

To realize the potential of a good concrete mix, placing must be done properly:

- Formwork must be robust to withstand vibration during compaction.
- The size of vibrators and their vibration frequency and duration of use must be properly controlled. Most vibration is done by motor-driven pokers moved by hand and immersed in the concrete as required, but vibration can be applied externally to the formwork if this has been designed to cope with the stress.
- Grout leakage must be minimized because its loss increases permeability.
- Placing and compaction must be done within specified time limits to ensure workability.
- Pouring should be done close to the final position. It is undesirable to work concrete along the shuttering because the mix tends to separate, causing layering which can increase permeability.
- Successive pours should follow one another reasonably quickly. If they do not, the effectiveness of interface bonding diminishes.

Concrete develops its eventual chemical structure over a long period, but the initial stages of curing are particularly important. The concrete must not get too hot or too cold or be allowed to dry too quickly. As remarked earlier, the water content is critical, and if it is reduced at the surface by premature drying, a poor set will result at the concrete face. Shrinkage cracking is then likely and this can propagate inward into the body of the mix.

Waterproofing additives and daywork joints

Waterproofing additives are available which inhibit the passage of moisture through concrete. However, it is usually the interface junctions between pours that leak, not the concrete and the use of additives will not prevent junctions leaking. Moreover, the soapiness of some of them will actually inhibit good bonding. There should always be a good reason for specifying an additive, for example to facilitate the pumping of concrete.

It is therefore the joints between pours on which one should focus attention for unprotected basements. Even if the shrinkage or reinforcing steel crosses these interfaces, and even if they are clean, the bond cannot be as strong in tension as the body concrete, so normal shrinkage will often permit some seepage by capillarity. Any settlement will increase this risk, and on most building sites the reality is that if there is much of a time lapse between pours, the face of the concrete already set will get dusty or dirty or be frost-damaged or weakened by quick drying and be unable to bond well.

Roughness assists bonding and if possible the pre-set surface should be treated with a bonding agent just before the next pour. Proprietary bonding agents are available, but I have also had success with a very early BRE recommendation of a watery solution of cement vigorously brushed on. The scrubbing dislodges loose material, the water prevents too much de-watering of the fresh concrete at the bond face, and the cement enriches the interface contact to give added strength. Note that the solution should be watery, not a creamy grout, and should go on more like a stain than a paint.

If it is not practicable to do these things, a joint cover can be applied externally, such as the one in Figure 3.4 which is an extruded rubber strip with undercut ribbing. It can be placed on a smooth prepared base where a floor joint is to be made or fixed to the wall shuttering at the point where a wall joint is to occur. End-to-end connections of the ribbed sheet must be tightly made or leakage will take place there.

Proprietary water bars of metal, rubber or plastic are also available, and these are intended to be buried in the depth of the joint (Figure 3.5). However, experience has revealed some problems. They can seldom be used where reinforcement is to cross the joint. Concrete does not bond to them, so that seepage by capillarity sometimes occurs, and end-to-end junctions and corners have been difficult to form reliably, so that leakage takes place.

Figure 3.4

Figure 3.5
Water bars are made in several shapes.

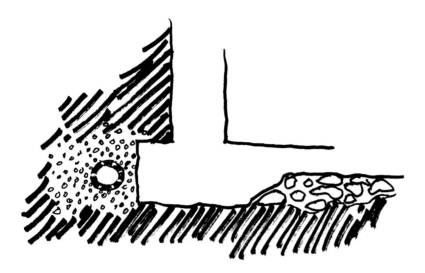

Figure 3.6
Untanked basements should have drainage.

Drainage

Single-thickness basement walls should always be provided with drainage around the footings, using open-jointed or perforated drain tiles bedded in gravel (Figure 3.6) and draining to some safe position nearby. Preferably the tiles should be covered by a filter fabric before the gravel is placed, and it may also pay to run one or two drains under basement floors where these are large.

Hardcore

Sub-floor hardcore is usually an open-textured mix of crushed brick and concrete or shale, tamped or rolled into place. It is sometimes used as a bed below a basement floor but more commonly is found below slabs at ground-level. The concept is traditional, but the benefits of using it need to be thought out in any particular case. For example, it has obvious value on clay soils where self-drainage is likely to be poor, but may be pointless on sandy soils with good self-drainage.

Soil should always be checked for sulphate content to avoid degradation of the concrete.

Moisture content

Concrete walls contain a lot of water initially, and how soon basement walls will be dry enough for decoration is uncertain. Drying can be accelerated a little externally by delaying backfilling, but what happens after that depends a lot on the natural moisture content of the contact soil. In the long term this will determine the moisture content of the contact concrete because the two will establish equilibrium at their interface. Then the dampness of the inner surface will depend on the vapour flow rate through the concrete and the indoor loss rate by evaporation and ventilation.

The same is true with the floor, but more formidably, and a word of warning must be sounded about floor coverings. Rubber-backed carpeting, cork, lino or any other such low-permeability floorings can only be used if the floor incorporates a damp-proof membrane to prevent upward vapour flow. It must be a true DPM, not a brush- or spray-applied material, because these are too vulnerable to poor application. Otherwise evaporation must be allowed to take place through the floor covering or a smelly green mess will form under it.

It can be taken as axiomatic that an untanked basement must have reasonable ventilation.

A summary on untanked basement design

The viability of an untanked basement is a matter of judgement for the designer and depends on the use intended for the space and whether the owner is prepared to accept the risks entailed. The owner's agreement should always be obtained in advance, in writing.

Seepage causes ugly staining and can be difficult to avoid or remedy. It can adversely affect the market value of a building, even where the basement is only to be used for car parking, because people do not expect buildings to leak. If electrical or electronic equipment is to be used in an unprotected basement it could be impaired by dampness and even become dangerous. Careful consideration should also be given to the storage of any delicate or moisture-sensitive material. Powered ventilation might be necessary, or even air conditioning.

3.2.4 Protected basements

Protection is achieved by the adoption of one or other of three strategies: the external enclosure of walls and floor by a water- and vapour-tight membrane, or by a double wall and floor system with such a membrane sandwiched between the two leaves, or by a cavity wall and floor, with floor drainage. With the first two strategies the floor membrane has usually to be laid in a sandwich because it is difficult or impossible to form a membrane reliably on soil under a main floor slab. It usually gets damaged before or while the concrete is placed and in any case it is then too difficult to bond the floor membrane to that of the walls.

Membranes

These strategies are discussed in detail below, but whichever is used, the strips of floor and wall membrane must be bonded together securely to form a water-tight enclosure. They should therefore be of similar material and detailed in a way that enables operatives to do the bonding reliably on-site, for if it is not done well, leaks are difficult or impossible to remedy.

The requirement is for a reinforced water- and vapour-tight sheet tanking material – typically a high-performance bituminous felt, well lapped and sealed below floor screeds and applied with a hot bitumen adhesive to walls, or a bitumen felt modified to be applied and lap-sealed by torching, though these usually require a primed base as well.

It is wise to double at least the floor membrane to remove any risk of lap leaks or damage during construction.

Membranes should never be directly adhered across the main casting joints in the concrete, for movement or shrinkage of the concrete is to be expected and may cause splits in the membrane along the joint lines. Slip strips 150 or 200 mm wide should be laid along them to prevent adjacent adhesion. Reinforced bituminous membranes will usually stretch reliably over honeycombing in the concrete or the minor shrinkage cracking that may develop, but coarse honeycombing should be filled in first.

Membranes should not be applied to concrete sooner than a week after it has been poured, and preferably not for two or more weeks. Otherwise, the water content of the concrete is likely to cause blistering or more general debonding.

Polythene has been used occasionally for tanking but it is difficult to manage in windy conditions and it cannot be fixed by adhesive. Laps have to be sealed by taping and corners have to be cut and taped, and these processes are vulnerable to careless site workmanship. Often the tape comes unstuck, too. Moisture appears to be able to pass gradually through polythene.

Asphalt, a material with a long and worthy track record, has the disadvantage for basements that it has no reinforcement and, compared with good reinforced felts, is therefore more liable to develop cracks if there is any movement of the concrete. If asphalt is to be used under heavy column or wall loads, manufacturers should be asked for advice; it can squeeze out under pressure. Some asphalts flow more readily than others for reasons explained in the main discussion in Chapter 9.

Numerous failures of brush- or spray-applied coatings have been seen. Dirt, dust and friable surface concrete are inevitable on building sites and militate against reliable adhesion and coverage. Also, these coatings cannot bridge over honeycombing or coarse porosity and do not lie well on rough surfaces because they thin out over sharp edges. Operatives also often thin them in cold weather to make them easier to apply and can overdo the thinning – sometimes deliberately.

Three-way corners are always vulnerable positions in membrane application. No matter how they are lapped on inward or outward turns there will be a no-lap point where the three planes meet and this is where water will get through. The solution is to specify pre-formed corners and good laps. See Chapter 4 for development of this advice.

Floor design

The main floor slab will usually be cast on a hardcore or prepared soil base and will carry the tanking membrane on which a screed will usually be laid. The main slab is usually reinforced and cast with any necessary footings for walls and columns. The slab may be subjected to upward bending stress due to hydrostatic pressure and/or expansion of the subsoil and/or relative settlement of the load-bearing elements as their loads develop. The reinforcement should have been designed to resist these forces and must continue across all daywork casting joints to unify the slab and distribute shrinkage stress. It should be finished smooth enough to provide a good surface for the membrane. If the basement is likely to be subject to significant hydrostatic pressure, the membrane should be doubled, with both layers fixed well by adhesive to await the screed. The concrete should be clean and dry to receive it and membranes should be lapped at least 100 mm, preferably more. Don't take risks in this position.

The floor and wall membranes have to be sealed together to provide tanking of integrity. When the wall is only a single thickness its protective membrane will be external, and the floor membrane must therefore pass under the wall to meet it (Figure 3.7). There must be sufficient external selvage, at least 150 or 200 mm, to allow a good bond to be made and a firm surface is needed on which to make it. This will usually be on the

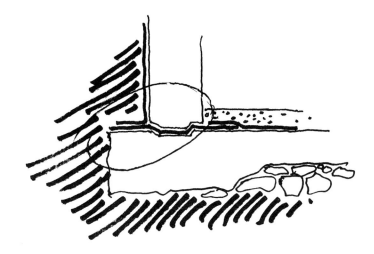

Figure 3.7
External tanking, floor-to-wall junction.

flat projecting edge of a footing. The selvage may have to be protected from damage during site works and be cleaned before bonding.

On no account should footings be cast separately from the main slab unless they are securely bonded by reinforcement. It would introduce a risk of shearing the membrane along the meeting line if differential settlement were to take place.

The screed should be reinforced and the reinforcement should be continuous across all its daywork joints to prevent them opening up due to shrinkage, for this could pull and damage the membrane beneath. The screed will also help keep the membrane laps sealed if or when hydro-static pressure develops. It may therefore need to be heavier than normal and could even be designed as a supplement to the floor structural system.

Changes in the plane of the floor slab due to stairs, ducts, lift bases and other such things should be avoided wherever possible, and when unavoidable they should be well thought out in advance so that site waterproofing operations can be kept simple. It is advisable to do 3D sketches to establish for oneself and for operatives the way the membrane is to be managed so that bonding laps can be truly sealed, especially around corners. All this is vital for reliable waterproofing.

Where the basement wall is a single thickness it should not be connected to the floor system by reinforcement, because this will prevent the floor membrane being taken through intact beneath the wall. Instead, a step or groove can be formed near the edge of the floor to help the walls take any inward thrust of soil and water pressure. The step should be formed with chamfered edges rather than right angles so that the floor membrane does not have to be bent into 90° turns. These increase the risk of cracking, especially in cold weather.

Do membranes get damaged by compression under wall or column loads? Bitumen in membranes is unable to flow any significant distance and mainly compresses into a denser layer. An advantage is that some of it gets squeezed into an intimate and more water-tight bond with the concrete. None of the non-bituminous membrane materials has this useful characteristic, and some of them contain a plasticizer which can be squeezed away from pressure points, causing areas of depletion where the pressure occurs and over-enrichment nearby, either of which can cost the material its integrity. See also Chapter 4.

If the floor membrane is only a single thickness it is desirable never-theless to double it under walls and columns. If the column loads are very high, consulting engineers sometimes advise a sheet overlay of copper or zinc or aluminium on the felt under the column. Its usefulness is not clear, but in any case it should not replace the membrane material; this must be continuous.

It is sometimes claimed for some non-bituminous membrane materials that they do not compress significantly under load. In fact, none of the waterproofing membranes will compress more than a millimetre or so at most, and this is irrelevant in the construction of buildings.

Working space around basements

The area for excavation has to be larger than the basement itself to allow space for placing and removing shuttering and there should be enough room for an external wall membrane to be applied properly and bonded to the floor membrane along its selvage. Backfilling should not be done until any specified drainage and gravel is in place around the footings, and lap seals where wall and floor membranes are bonded have been checked. If there is not enough space to do this important job properly, its reliability must be suspect.

The main area of floor membrane must not be left exposed to damage by indoor site works. In the single-wall strategy this means either that the areas of membrane to go beneath the walls should be put in place ahead of the open areas with a good inward- as well as outward-bonding selvage left exposed, or that the membrane should be laid as a whole and its exposed areas protected.

External protection for walls is advantaged naturally because soil and water pressures act positively to keep a membrane in place. Protection applied internally risks being dislodged by vapour or water leakage pressures exerted inwards through the wall.

On wettish sites it may be sensible to apply a protective sacrificial membrane over the external wall membrane and/or to double the latter to increase reliability and extend the life of the waterproofing, although in practice bituminous products below ground, where they are not affected by ultraviolet and ozone, usually survive well. Where there is a risk of mechanical damage from site works, a single leaf of masonry is sometimes built against the membrane (Figure 3.8).

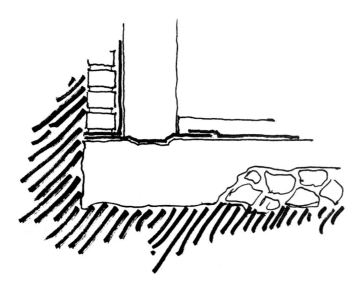

Figure 3.8
The external membrane protected.

With external protection a decision has to be made about the handling of the top edge of the membrane. It must come up to at least ground-level, but if it is exposed above ground it will not only look unattractive but will be at risk of damage and its life will be attenuated. If a protective leaf of brickwork is applied over the membrane below ground, the transition to the damp-proofing of the wall system above ground should present no difficulty, but otherwise some careful thought needs to be given to the way this potentially awkward position should be handled architecturally.

Workmanship must always be well checked on basement waterproofing: repairs are difficult, sometimes impracticable, and always expensive.

The sandwich wall option

The point has been made that a membrane applied internally to a basement wall is at risk of being forced off by water or vapour migrating inward under pressure through the concrete or masonry, so the established and sensible convention is to sandwich it between the outer wall and an inner leaf.

Compared to the single wall with external membrane, the use of an internal membrane requires a different wall construction sequence. Instead of building the outer wall on a DPC, it would be built directly off the concrete of the floor slab and the two would be connected by reinforcement but the inner leaf would be built off the edge of the floor DPM and this would be turned up the inner face of the outer wall to join its membrane. The membrane and the inner leaf would then go up more or less together, with mortar fill to keep the membrane tight against the outer wall. Severe soil wetness would justify doubling the membrane (Figure 3.9).

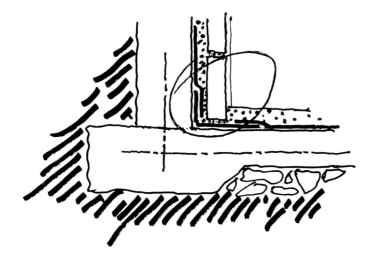

Figure 3.9
An internal membrane sandwiched.

The cavity option

If the cavity wall and floor option is adopted, the outer leaf will be a normal self-stable basement wall. Any water getting through would be expected to collect at the bottom of the cavity, as usual in cavity walls, but instead of being discharged externally by a DPC, it would simply be allowed to pass beneath a reinforced raised screed and be led thence to a sump from which it would be pumped away or escape by gravity. The

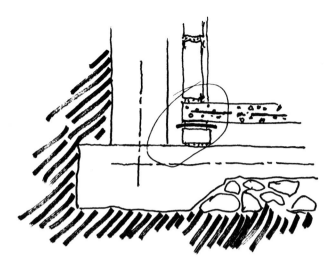

Figure 3.10
The cavity option: leakage water to go to a sump.

raised screed would be on supports with DPCs which would insulate it from moisture. The inner leaf of the wall might simply be a visual screen to keep the main basement wall from being seen and it could be supported on the edge of the screed or in any other way appropriate to its construction (Figure 3.10).

Where masonry construction is chosen for the inner leaf, the usual care must be taken to prevent mortar droppings collecting at the bottom, and the cavity must be ventilated, whether to indoor air or from inside to outside via vents above ground-level is an open question.

3.2.5   Treating existing unprotected basements

I have sometimes been asked to reduce infiltration through the brickwork walls of oldish basements. The mere existence of infiltration tells one that there is enough dampness in the soil to create some hydrostatic pressure. It may simply be diffusing inward but it could be getting through some cracks as well. Either way it is unlikely that any internal surface will avoid being stained, and with cracks present some continuing leakage is likely. The only safe course, as it seems to me, is a treatment along the lines of the sandwich wall option described above, laying a DPM on the floor and bonding it to a DPM applied internally to the wall, and then screeding the floor and forming an inner leaf for the walls to keep the DPM securely in place.

3.2.6   Pipes and conduits

Pipes sometimes have to pass through basement walls and there will be occasions when soil dampness creates a leakage risk. Pipes of clay or cast iron are sometimes cast in with the concrete, but it is then probable that the concrete will not get consolidated fully around the pipes or will shrink away, leaving a capillary path.

In principle, a sensible technique seems to be to cast an oversize removable sleeve in the concrete (Figure 3.11) and then caulk around the pipe from inside when it is in place. A collar of some kind has to be fixed on the pipe at the outer end of the opening so that there is

Figure 3.11

something against which the caulking can be pressed for consolidation. This allows the caulking to be renewed from inside as necessary. The services engineer should be consulted.

3.2.7 Thermal insulation

Dry soils provide reasonable or good thermal insulation, damp soils only a little, and wet soils are conductors. If it is necessary to improve the natural thermal conditions indoors, externally applied insulants such as foamed glass or extruded polystyrene are possibilities, for these are reasonably water-repellent or waterproof, and the outside is the best location for insulants. They would need protection against mechanical damage, presumably by some form of outer leaf as with the external application of DPMs described above.

3.2.8 Multi-storey basements

In principle, multi-storey basements will normally be cavity or membrane-protected sandwich systems, but the inner wall may be formed as part of the internal structural system and the outer wall built to serve as a retaining wall, at least until the internal structure can add its support. However, these are matters for the consulting structural engineer and the architect to decide together.

3.2.9 Underground buildings

Basements can be extended under courtyards or gardens and used for storage or parking or even offices. Complete buildings can also be sited below ground, such as shopping malls, offices, schools, and even homes. Sometimes the aim is to take advantage of the high thermal insulation that dry soil can provide but sometimes it will be to exploit expensive land.

All such uses require total enclosure by a waterproofing membrane system, bonded integrally throughout. The most sensible course would appear to be to treat the bottom level and the walls along the lines described above for the design option using internally applied membranes and to treat the roof along the lines to be described in the later discussion of flat roofs carrying gardens (Chapter 9). The perimeter of the roof presents a special problem because the wall/roof junction cannot be protected by overhangs or other conventions used for exposed roofs.

The slab supporting an underground roof is best cast *in situ* for if precast slabs are used, there is a risk of movement at joints which might damage the membrane.

When patios, courtyards and light wells are dropped into underground buildings, parapet construction of some kind is necessary. In principle, the situation is akin to a parapet edging to a gardened roof and is discussed in Chapter 9 in the context of flat roofs.

Case note

A patio/light well was dropped one storey down in a two-storey underground building and its walls were glazed full height. A brickwork parapet around the edge of the wall had a faulty DPC arrangement that let water through to the tops of the metal window frames, formed in a U-shape like a gutter. The vertical window mullions then acted as downpipes to deliver water down to cill level whence it spread indoors over the floors.

Even if an underground building is simply to be used for parking, the quality of overhead protection should not be downgraded. Leaking water will become alkaline as it passes through concrete, and alkaline drips from a roof can damage the metalwork of suspended ceilings or equipment and damage some types of paint on cars. Liability for such damage may be assigned to the designer if the membrane specification was inadequate, or to the contractor in the case of bad workmanship.

Case note

A large underground car park had a garden over it. The roof under the earth was simply painted with a so-called 'waterproofing' paint. An enormous quantity of alkali-bearing water leaked through and did considerable damage to cars and pipework. The garden, together with its large fountain, had to be removed and a proper membrane protection put in place (Figure 3.12). If cars are to be parked directly on the roof of an underground building, heavy-duty waterproofing is needed (see Chapter 9).

Figure 3.12
Part of a basement ceiling under a gardened courtyard. The water protection was a thinned proprietary 'waterproof' paint. Water went through, became alkaline during seepage through the concrete, formed stalactites and dripped onto pipes and parked cars, causing damage to paintwork.

3.2.10   Site problems during basement construction

If substantial sites have been properly sampled in advance, unexpected problems with water do not often occur. On risky sites engineers will probably have recommended de-watering beforehand by one or other of the established techniques, and if problems occur, their management will fall mainly to the consulting engineers. If troublesome amounts of water do appear, they may have to be taken to a temporary sump by pipe or gravel water-course and cleared there by pump or gravity to a storm water main or other safe discharge position. It is important to ensure that this water does not remove soil that could destabilize an adjacent building, and it is also necessary to avoid discharging mud and silt into the storm drainage. Filtering may be necessary.

Limestone and clay soils

Some limestones and clays deteriorate and soften when newly exposed to air and moisture. Protection may have to be provided until construction begins and can often best be done by a layer of weak concrete, applied as soon as possible over the exposed areas.

The interface between new and existing buildings

Where a new building interfaces with existing buildings, it is usually important to find out how the latter are founded so as not to expose them to any settlement or waterproofing risk and also to avoid unexpected problems when designing the basement of the new building. If possible, obtain drawings and other information about the adjacent foundations or basements in advance and, if necessary, check their reliability when site works begin.

Protection of materials

Wet sites can be a chaotic morass where operatives become careless and put themselves and their equipment at hazard. Poor workmanship and slow progress are likely results. Good site management is needed in such circumstances, otherwise materials may be inadequately protected and deteriorate or be used in an excessively damp state.

3.2.11   Water tables

Water tables which are low on initial checking can later become high due to seasonal change, swings from periods of drought to periods of rain, or as the result of some happening nearby. Burst water mains and leaking sewers are becoming an increasing hazard that can also raise the local water table.

Modern site de-watering technology makes it possible to build virtually any kind of basement on a site with a high water table, but specialized design may be needed. For example, floatation is a risk if the building is not heavy enough to counter the highest levels, and an upward overflow pipe system is needed to release the pressure at some predetermined level of the water table. Such an arrangement would call for a drainage blanket under the bottom of the basement to allow the water to run easily into the overflow pipes, and thence into sumps to be pumped away. It is the kind of thing that requires good architect/engineer collaboration during design.

3.2.12   Design responsibilities

As in a few other areas of building design the responsibility for dampness protection in basements falls sometimes to architects, sometimes to collaboration between them and the consulting engineers and sometimes almost or entirely to the latter. The boundary line between the professional liabilities has no fixed position and depends on case-by-case circumstances. A confusing element in locating it is the common belief that because architects are understood to have the general responsibility for waterproofing the building envelope above ground they also have it below ground, but it should be clear from what has been said here that this cannot logically always be the case.

In practice it is unusual to attempt to define the boundary position in the euphoria of the early stages of design, but when a basement leaks such euphoria quickly evaporates as the client begins to consider who might be to blame. Usually only a general demarcation of responsibility can be made in advance, but as the design develops the issues clarify, and their sharper identification may point to the need for more formal agreements before any trouble occurs. This is a different matter from the risk of leakage due to faulty workmanship for this will involve the contractor, although it may also bring in whoever had responsibility for inspection.

Basements are accident-prone and their buildability should be discussed with the contractors before they begin work.

3.3   Works at ground-level

The American term 'slabs on grade' is adopted here as a useful abbreviation for 'laid directly on the ground' without a basement.

3.3.1   Slabs on grade

The convention for floor slabs on grade is to remove the top soil, put some hardcore in its place, and then cast about 100 mm of concrete on it, covering this with a DPM and finishing it off eventually with a screed. Hardcore is usually the crushed remains of demolished masonry buildings and then consists mainly of crushed bricks, mortar and concrete. Shale, gravel and other fills are also often used when locally available. Some fills contain salts which can cause troublesome chemical reactions to metals or concrete while other fills may expand and cause heaving. If waste from tips and other dubious sources is to be used, make sure that it is chemically harmless.

The use of hardcore is a long-established tradition and traditions that survive usually do so for good reasons. However, the functions that hardcore performs are not clear. On wet and heavy clay soils it no doubt helps to insulate the slab from some of the wetness, and it will reduce the thermal capacity of the floor system a little as compared with a slab directly laid on damp clay, but none of this seems to have been researched.

Hardcore cannot be precisely specified and there may be as much clay as rubble in it. Sometimes it is 'blinded' by a finer top dressing of some kind or protected by polythene to prevent the slab concrete from losing particulate material and the water it needs for a proper set. On sandy, well-drained soils hardcore seems hardly necessary if the soil itself is stable.

While hardcore can insulate a slab from contact with wet subsoil, in an oceanic climate it is to be expected that water vapour will transpire from the ground even if it is not visibly wet; this can maintain a significant moisture content in the slab concrete. Whether this matters or not depends on the uses to which the indoor space above may be put. If it is to be for garaging, for example, the slab can be self-finished and evaporation may then cope with the moisture, but if the space is to be used as a dwelling or office, the floor system should incorporate a proper DPM.

Case note

A sports hall floor of 12 mm plywood on sponge-bottomed battens was laid on a thick screed over a polythene DPM on a main slab. The site was wet clay and no hardcore was used (Figure 3.13). The screed became damp, the plywood gradually did likewise and expanded, closing 10 mm ventilation gaps that had been left around the perimeter. Warping also developed and the floor became too dangerous for sports use.

Moisture meter readings unexpectedly showed that the moisture content of the screed above the polythene had become the same as in the concrete below it. Evidently the wetness of the site required a membrane of much lower permeability, e.g. a high-performance roofing felt, perhaps even doubled. The gaps around the edge of this large sports floor could not provide anything like enough ventilation to remove the evaporating water vapour.

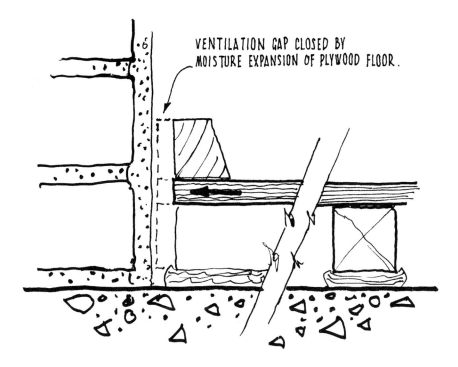

Figure 3.13

In such cases time is a factor, for it took several years for the problem to develop, but the site was known to be very wet and it should have been a take-no-risk decision. If it is necessary to use a DPM, it should always be the equivalent of at least one layer of good roofing felt, not merely a vapour check.

DPMs have two further functional requirements to satisfy for slabs on grade for low-rise buildings. First, they need to be able to resist splitting or tearing. When such a masonry building is under construction the wall and column loads gradually increase, and on clay soils the footings will settle a little, perhaps 10 or 12 mm. If the slab has been cast together with the foundations of the walls, the slab will take some of the load, but it will be bent down a little by the settlement. Although green concrete often bends quite well, cracks do sometimes occur and the DPM should be strong enough to cope with this. If it is, it will doubtless be able to maintain its integrity over honeycombing, poorly tamped finishes and other such shortcomings. The second requirement is that the material used for it should be the same as that used for the damp-proof courses in the walls of the building so that they can be bonded together reliably.

3.4 Suspended timber floors

The suspended timber floor traditionally used in domestic design in Britain was abandoned after 1945 to avoid the need to import timber. It has gradually come back into limited use, but now needs thermal insulation. Standards for this are revised periodically and current regulations should be consulted.

The convention is that concrete is laid over the ground, without hardcore. There is then a ventilated void over which an insulated floor is carried on sleeper walls built off the slab and perforated, usually by open perpends in brick- or block-work, so as not to preclude cross-ventilation in the void.

The assumption that cross-ventilation is the mechanism by which moisture in the void is removed has led to regulation requirements for openings at a specified rate per metre along its exposed perimeter. However, for cross-flow to take place there must be air pressure differentials externally on at least two edges of the void and these will often not exist. In fact, although the matter has not been researched, the avoidance of high humidity in the void must also depend on the dispersal of high vapour pressure levels by their normal search for equilibrium with the external vapour pressures, as described in Chapter 2. The two mechanisms require in principle similar provision for pressure cross-flow but for different reasons. Where the voids are much larger in area than those for conventional domestic buildings a risk exists that ventilation by cross-flow will not be effective, nor will high vapour pressures necessarily be able to disperse if reliance is placed on the norms used for houses. I have seen one or two cases which suggest that research could provide improved guidance for extensive voids. Meanwhile, a good rule would be that the larger the void area, the more generous should be the provisions for vapour to escape.

It is not usual to introduce a vapour barrier in the suspended construction unless this is required for radon protection. Refer to the BRE Guide on the subject.[1]

3.5   Paving

Where concrete paving slabs are laid on buildings or their balconies they should not be bedded on a lime:sand mix as is commonly done for pedestrian pavements along streets. The lime leaches out into the rain run-off and when this very alkaline water meets the air it reacts to form calcium carbonate deposits, usually at drain entries or the mouths of the downpipes. It can soon build up into a blockage, sometimes complete, and not easily removed. Set slabs on sand and trap it to prevent it being washed away.

Cyclic thermal movement takes place in areas of paving slabs which causes their gradual collective expansion. The mechanics are simple; in warmth the whole area expands a trifle but on cooling the slabs shrink individually, leaving fine interface gaps. Dirt gets into these and some of it is hard, so that the next cycle of expansion is slightly greater overall. The collective growth eventually takes place with considerable power, sufficient to damage asphalt skirtings around the perimeter of a paved area or even to cause cracking in brickwork. It happens regardless of whether the slabs are butt-jointed or have mortar jointing, and allowance should always be made for it by soft edging and by re-laying every few years.

Clay brickwork has not proved very successful for outdoor paving. Water gets in at joints so that the bricks stay wet on the underside and this creates maximal long-term moisture expansion, to which solar heating will also contribute. Cases have been seen where this has been so disruptive that the paving became unreliable for foot traffic and had to be re-laid. Exposed to the weather, the wetness of the bricks also made them vulnerable to frost damage. Soft joints simply allow more freedom for the movement and lead to damage to high-heeled shoes or umbrella ferrules. The disruption also looks very messy. The more recent convention of laying concrete bricks dry on a sand bed has proved more successful, presumably because there is no long-term build-up of moisture expansion. No cases of collective thermal expansion are known but the reason is not clear.

3.6 Simple foundations for low-rise construction on clay

Before 1939, wall foundations in Britain for low-rise brick-built property of two or two-and-a-half-storeys had become established by habit (and later by by-laws) with footings 2 ft 6 in. wide and 2 ft 6 in. deep. The width had nothing to do with the load to be carried: it was simply a convenient working width for a man to dig them out, and the depth had nothing to do with the soil shrinkage zone: it was what experience had shown to be about right for finding undisturbed bearing soil in most parts of the country.

Settlement failures occurred but were not researched; they were simply rectified in whatever way seemed sensible to the local builder. The generally prevailing view, even in engineering circles, was that what happened below ground was in the lap of the gods.

However, change eventually took place. Geotechnic engineering was in active development at the Building Research Station in the late 1930s, and given the priority that house construction enjoyed post-1945, some attention was inevitably directed to house foundations. It was realized that the loads of brick-built houses were too small to cause settlement by overload and that when it occurred, it was usually due to withdrawal of support by drying shrinkage of the subsoil and/or its de-watering by the roots of trees or large shrubs. These studies pointed the way towards strip footings 9 or 10 in. wide and cut by machine to a depth of 3 ft or more.

It was seen that the problem was largely related to clay soils. Clay prevails in Britain in areas roughly south of a line between the Mersey and the Wash. As described earlier, all kinds of clay shrink when de-watered and swell on re-wetting, both processes taking several months or even two or three dry or wet seasons to develop fully. It became clear therefore that the loads had to be taken below the shrinkage zone. Tests with short-bored piles in common clay soil indicated that 10-in. bores taken down 6 or 7 ft and set at intervals of 6–8 ft or so around the walls, and doubled under point loads such as chimneys, should be sufficient to cope with the loads involved.[3] The bores could be cut by hand with a common posthole augur (Figure 3.14).

It was also realized that the beam needed to carry brick or block walls between the pile caps would not have to be very substantial because in masonry the loads arch between points of support. The suggested design basis was WL/250, now revised to WL/100, which could be met by a lightly reinforced thickening of the slab edge to about 9 inches, since it had little more than its own weight to carry. I have usually used two 12 mm bars near the bottom and one near the top. It followed then that piles would be needed either side of an opening which went down to slab level because of the interruption of the masonry.

I used this approach on my own two-and-a-half-storey brick-built house designed in 1946 (Figure 3.15) and have used it on several hundreds since then. I can offer the following observations.

Usually the piles are easily cut in an hour or so down to a depth of 3 m, which gives an added safety factor, but if there are enough to do, it pays to cut by tractor-mounted power augurs – a cut then takes only two or three minutes. If piles have to be left overnight before filling, a little water sometimes gets in and softens the toes, and these should be

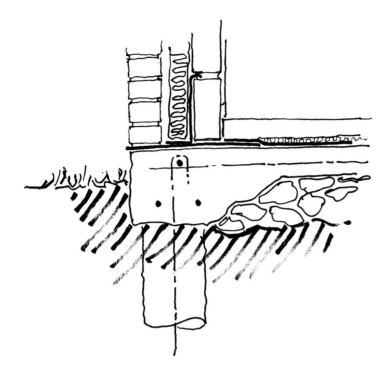

Figure 3.14

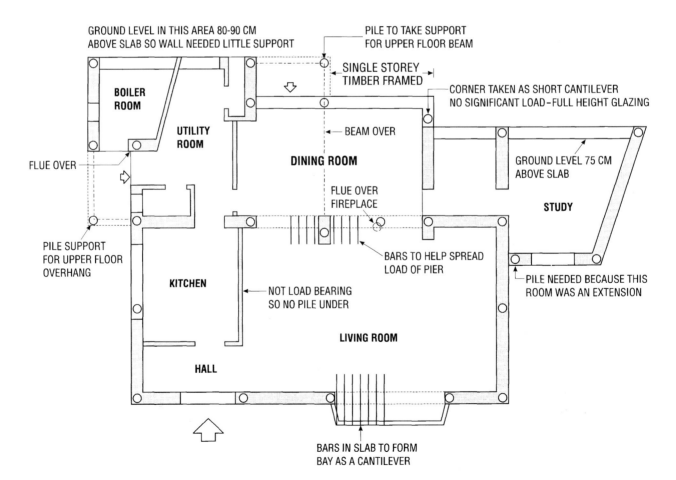

Figure 3.15
Plan of the author's house, 1946, showing pile locations.

cut a little further down. If the soil has biggish stones or roots, a 300 mm diameter augur will usually speed up the work.

The piles as I have used them were reinforced with a single 10 or 12 mm rod hooked at the top to engage the beam reinforcement and this will also remove any uplift risk on the beam before the wall loads are in place. The beam described above seems to have been entirely satisfactory. On sloping sites, and perhaps generally, it is sensible to reinforce the slab. Rods of about 6 mm in a 600 mm grid seem to have been successful.

Instead of taking the foundations around such projections as bay windows, it is often simpler to cantilever the slab out over the beam, which continues beneath it across the gap, if they do not project too far (Figure 3.15).

Short-bored pile foundations involve the removal of only 25–30 per cent of the soil taken out for conventional footings and requires correspondingly smaller amounts of concrete, with, in my experience, a useful reduction in costs. In addition, the resulting cleanliness of the site gives the secondary advantages of general tidiness and quicker work. Tree proximity to about 5 or 6 m seems to be acceptable except for Lombardy poplars, which have an exceptional spread of roots. In general, it is possible to build safely with piles fairly close to large trees (Figure 3.16). In

Figure 3.16
This was a small, but promising, oak when this house in Welwyn Garden City was built in 1951. The house is on piles 8 feet × 10 inches, and growth of the oak in the 55 years to the most recent inspection has revealed no signs of foundation disturbance.

fact, I have placed piles successfully as close as 2 m distant from young trees, but always then using pile depths of 3 m. Caution is desirable with all foundations, as well as judgement based on local knowledge of soil conditions.

Houses on sloping sites always tend to drift downhill slightly. Piles cannot stop this any better than strip footings do, but I sometimes wonder whether tilting the piles would resist it. A tilt is easy to do.

In my view, this sort of foundation is a desirable convention to establish for houses and other low-rise property on clay subsoils in Britain in the interests of improved reliability, economy, site cleanliness and the reduced use of concrete.

I have also been able to adapt the technique for inexpensive remedial work to older foundations which have settled or become restless. In many cases dry-weather subsidence is largely or fully rectified when the soil re-wets, usually over the winter season, and early spring is the time to do such underpinning. This is then done by cutting piles close to the wall line and forming reinforced shoulders under the wall to get its load over to the pile caps.

It has never been the general practice to make house foundations a matter for engineering design but some Building Control officers expect this as soon as piling is proposed, despite its inherently greater reliability in the form described. It has been my experience that overdesign sometimes then takes place and the potential economy is eroded. For example, I have seen three 6 m piles proposed for a single-storey extension only 1 m × 4 m on plan and consisting mainly of windows.

References

1   BRE Digest 263, Sulphate and Acid Resistance of Concrete in the Ground, HMSO, 1991.
2   BRE Report, Radon: Guidance on Protective Measures for New Dwellings, BRE, 1992, £8.
3   BRE Digest 63, Soils and Foundations: Part 1, HMS0, 1965.

Further reading

1   CIRIA (The Construction Industry Research & Information Association), Report 139, *A Guide to Water-Resisting Basement Construction*, 1995.
2   BRE Digest 348, Site Investigation for Low Rise Building: The Walk-over Survey, 1989. Also Digest 381, Continuing Site Investigation for Low-rise Property.
3   Davey, N., 'Short concrete piles for foundations on shrinkable clays', *J. RIBA*, **57**, 24–25, 1949.
4   BRE Digest No. 3, House Foundations on Shrinkable Clays, HMS0, 1949.
5   BRE Digest 42, The Short-bored Pile Foundation, HMS0, 1952.
6   BRE Digest 240, Low Rise Buildings on Shrinkable Clay Soils (in three parts), 1976.
7   BRE Digest 298, The Influence of Trees on House Foundations in Clay Soils, 1985.
8   Ward, W. H. and Green, H., 'House foundations, the short-bored pile'. This was a paper to the Public Works & Municipal Services Congress

of November 1952, by two authors in the Geotechnics Division of BRE. It is the most definitive engineering paper on short-bored piles. Published in *J. Inst CE*, and also available from the Document Supply Centre of the British Library at Boston Spa, Wetherby, West Yorkshire LS23 7BQ.

9   *Subsidence of Low Rise Buildings*, a publication by the Institute of Structural Engineers, 1994.

10 *Radon Sumps*, a BRE Guide to radon remedial measures in existing dwellings, 1992.

11 BRE Digest 365, Soakaway Design. 1991. This deals with storm water from buildings on paved areas.

12 See also CIRIA publications on *Basement Waterproofing*, *Building on Contaminated Land*, and *Building on Fill*.

4 Cavity wall systems

4.1 Development of the cavity concept

The solid walls of which buildings were traditionally built mostly dealt with rain by shedding water or by absorption and subsequent evaporation in some combination. The walls typically had as thermal properties a high inertia and whatever insulation happened to result from their thickness and the materials of which they were built. Risks of condensation were minimal because the buildings they enclosed were usually well ventilated by fireplaces and air leakage.

When the construction was of solid load-bearing brick or stone, thermal inertia was often an unsatisfactory characteristic. In winter it made buildings slow to respond to intermittent inputs of heat, although this was offset sometimes and to some extent by internal linings of low thermal capacity – panelling perhaps, or wood lath and plaster spaced or 'firred' out on battens, or occasionally tapestries in grand buildings. In summer, of course, the high thermal inertia could be welcome for the coolness it provided.

Solid walls began to lose their validity for substantial buildings as framing developed and took over the load-support role. Weight became an encumbrance and wall thickness was quickly reduced to what were considered safe construction limits. However, as walls became thinner, the loss of rain protection and insulation began to be noticeable and cavity systems were given spasmodic thought.

The earliest cavity arrangement to become popular was not for framed buildings, however. House walls shrank from 13.5 to 9 in. solid construction in the drive for economy after the First World War, and this proved to be a turning point because rain penetration became unacceptably frequent and the dwellings needed more heat than many people could afford.

Something had to be done. At the time, rain penetration was the more pressing problem and in the 1920s this was dealt with by dividing the walls into two leaves separated by a cavity 2 in. wide, which became the norm for the 1930s. This also gave some gain in insulation but not much, and when house-building resumed in 1945 a determined drive was mounted to improve thermal comfort.

The first step taken was to replace the brickwork of the inner leaf by various kinds of insulating blockwork, and this was soon accompanied by a drive to get rid of surplus fireplaces and to improve the efficiency of those that remained, as described in the discussion of indoor climates. Eventually, whole-house heating became general, while a growing awareness of the need to save energy led to requirements for further insulation. A home had to be found for this, and cavities provided it.

The cavity was evidently here to stay, and when multi-storey framed construction was resumed in the 1950s, brick-and-block cavity walls became one of the established ways of cladding it.

In the early 1960s, intensive official encouragement of industrialized construction for medium-rise flats brought pre-cast reinforced concrete cladding on to the scene, and this heralded a flowering of innovation in cladding materials and techniques, improved concrete finishes, thin stone claddings, curtain walls, renderings and, later, thin sheet materials in rainscreen form.

The cavity principle was used with almost all of them, an exception being curtain walls, and the width of cavities has gradually increased to receive the larger amounts of insulation demanded as the need to save energy gradually intensified. And in its turn the increased insulation has encouraged a new view to be taken of the roles played by the four elements of cavity walls, the inner and outer leaves, the cavity and the ties. An awareness developed that with the inner leaf protected by so much insulation, it could become an active participant in the thermal dynamics of the interior climate, while the outer leaf's role could diminish to that of a screen against the weather and a shelter for the insulation.

Later, in Chapter 9, it will be seen that much the same idea has developed in respect of flat roofs, whereby the weather membrane and deck are protected by insulation which in turn has an overscreen, usually of gravel or paving, to shelter it from ultra-violet light and to keep it in place.

In this way, then, it has become possible for the whole of the main structure of a building to be protected so as to enjoy thermal and dimensional stability to an extent not previously known. The need for structural movement joints is reduced or removed; the thermal restlessness of the fabric, which causes much incidental cracking, largely disappears; thermal staining of indoor surfaces is minimized, and the thermal mass of the inner elements of the envelope can be exploited advantageously in heating and cooling strategies for buildings. This is a very important group of ideas for the future.

For all these reasons cavity wall systems are discussed in this one enlarged multi-part chapter. The first part will deal with what happens in the cavity itself, which has never been properly studied, and the remainder with various types of outer and inner cladding.

4.2 The cavity

4.2.1 Leakage and insulation

The earliest cavity was a nominal 2-in. gap, now 50 mm, probably chosen as the smallest size that would safely avoid being bridged by mortar squeezed into it from brickwork joints in the two leaves. It was also a convenient size in which to lift a batten in the cavity to catch mortar droppings which would otherwise accumulate to bridge the cavity at the bottom where they would become saturated and wet the inner leaf.

It must also have been evident that if 9-in. brickwork was prone to leakage, as it had proved to be, then a 4.5-in. thickness would be worse, and it was probably for that reason that ventilation of the cavity was thought necessary to help keep it dry. Air bricks were therefore required top and bottom with air changes presumed to be induced by the stack power in the cavity created by the relative warmth of the inner leaf in winter and the outer leaf in summer. By this means the whole system, outer leaf, cavity and inner leaf, could be kept reasonably dry and the cavity also provided a means of escape for indoor water vapour migrating into it through the inner leaf.

This picture was not so clearly seen in the 1950s when standards of insulation first reached an amount requiring use of the cavity, and the view was taken then that because cavity ventilation would have a cooling effect in winter and therefore in theory would reduce the energy saving provided by the insulation, the requirement for air bricks should be allowed to lapse. Since nobody knew exactly what was going on about dampness in cavities anyway, there was no argument.

Also, it opened the way for insulants of various types to be used, some of which filled the cavity. One type was pelletized and could be poured or blown, both into existing cavities and into new walls; another was a waterproof plastic that could be foamed *in situ* in the cavity. The pellet insulants were not wholly successful since they could tumble into cavities for sound insulation in party walls or pour embarrassingly through unnoticed holes into cupboards and ducts. Although they allow a limited airflow upwards and do not often transfer dampness across the cavity, they can be wetted if there are leaks in parapets or if spillage takes place off DPCs over the lintels. This can then wet the inner leaf and give an impression of leakage. The foamed plastic was a major mistake for *in situ* it sometimes shrank and split, creating passages by which rain could and did cross the cavity.

Now there is general acceptance that it is wise in new construction for insulants in cavities to be used in batt or slab form and kept from contact with the outer leaf, leaving a gap of 25 mm, or preferably more, with provision for the air in them to move through. The pelletized insulants which can be blown into the cavities of existing buildings fulfil a practical need, but the cavities need to be checked beforehand for risks of solid bridging or unnoticed openings in the inner leaf. If cavity party walls have been used, there is no way to close them off so it must be accepted that some of the insulant will find its way into them.

Insulants in cavities will not be as dry *in situ* as in the test rigs where their insulation values were established for certification. Cavity air will have an absolute humidity conforming closely to that of the external atmosphere, and in practice this will lower the insulation value by an uncertain amount unless the insulant is impermeable – and there are only one or two of these. Field data are confirming that in practice insulants are not giving the values used in computations based on certified coefficients, and a systematic basis for allowances is needed.

4.2.2 Thermal and vapour implications

The placing of insulation in the cavity causes the inner leaf and the support structure for the building to stay much closer to the indoor temperature, while the outer leaf and the ventilation air in the gap beside it follow more closely the outdoor air temperature and radiation conditions. How closely the outer leaf does this then depends on its own thermal inertia, the cavity ventilation, and the back-up insulation in the cavity. In outer leaves of brickwork and concrete, temperatures change relatively slowly and therefore only approach their highest and lowest levels in longish periods of hot or cold weather. Thin stone may need only a few hours to do this, and thin sheet metal or plastic only a few minutes. During their lifetime, therefore, they will all sometimes experience their potential temperature extremes, but the quicker their response, the more often they will do so, and the greater the insulation value in the cavity, the quicker and greater these responses will be. Condensate will therefore form more frequently and generously on the inner face of the outer leaf than in uninsulated cavities.

The sources of water vapour in a cavity will be rain-dampness in the outer leaf, the infiltrated outdoor air, and any indoor vapour exfiltrating through the inner leaf and insulation. Whether the inner leaf should be made air- and vapour-tight depends on the indoor requirements, as discussed in Chapter 2. Buildings which need highly controlled indoor air and humidity will have to be air conditioned and will then need the inner leaf to be sealed against both air movement and vapour migration, while interiors which are not at least power-ventilated will need to lose excess water vapour passively, and their inner leaf and cavity insulation should be vapour-permeable. Cavity ventilation then facilitates its escape to atmosphere.

There is one exception to this generalization – the open-jointed external rainscreen cladding to be discussed later (see Section 4.7.1). The entry of rainwater through its open joints is restricted by the technique known as wind pressure equalization in the cavity, and for this to occur quickly the cavity pressure build-up must not be able to dissipate indoors through the inner leaf. It must therefore be reasonably air-tight.

These last two matters are described at more length later when we come to the discussions of rainscreens and the inner leaf (Section 4.7).

Assuming that vents to atmosphere have been provided in the outer leaf, stack power to move the air layer in a cavity will depend mainly upon the temperature of the outer leaf because the insulation in the cavity will limit the influence of the inner. When the outer leaf is warm or hot the stack power will move the air strongly upwards, while on cool or cold nights it will stagnate or move downwards. Either way it will ensure that the cavity air is always on the move and not far different in temperature or humidity from the outer air.

If the insulants are freely permeable to air, they will then need some sort of moisture protection to reduce their loss of efficiency. This should be vapour-permeable if the building concerned needs to lose vapour, but it should provide sufficient screening to keep the ventilation air from moving too easily in and out of the insulant. It needs to function somewhat like a breather paper.

4.2.3 Firebreaks

Firebreaks are necessary to prevent fire flow up cavities. Barriers will usually be needed at storey heights vertically and intervals of perhaps 7 or 8 m laterally. If the stack effect is not interrupted, fire flow can become rapid and the temperatures intense, sufficient to turn even low-combustibility insulants into fuel.

The blocking of cavities for this purpose implies that each sub-division created must be ventilated separately, and the vents at the bottom of one cavity should not be placed immediately above the outlets from the cavity below, so as to avoid fire transfer.

The reveals around window and door openings must always be able to prevent fire in a cavity from breaking through regardless of whether the windows or doors are set outside or inside the cavity line; otherwise dangerous fire transfer can and does occur.

4.2.4   Thermal pumping

The five categories of outer cladding to be discussed here all use different jointing with differing permeability to external air. The open joint systems of rainscreens give free access, concrete and thin stone claddings usually have mastic seals and may be almost air- and vapour-tight, while renderings have no joints at all. Although they look air-tight, brick and blockwork joints have significant air and vapour leakage by micro or macro leaks even if air bricks are not inserted.

It is important to avoid creating cavities which are almost but not perfectly sealed, for these produce classic thermal pumping conditions (see Chapter 1). The air temperature in the void will always be changing with the weather, and as the air temperature changes so will the void pressure if the void is sealed or nearly sealed. The pressure will rise as the temperature rises and will drop as it cools. When the seal is imperfect, as it must usually be in buildings, air and water vapour will be drawn in through leaks either in the inner or outer leaves or around door and window reveals, and this will raise humidity in the cavity. Some of the vapour will deposit in the insulation, some will be absorbed into inner and outer leaf materials, and some will sometimes condense on the latter. Not all of it will revaporize easily and a gradual accumulation of dampness is likely, especially in winter. Outer claddings with mastic or mortar joints or no joints at all must therefore be provided with vents to prevent the development of a one-way build-up of cavity moisture.

4.2.5   Connectors across the cavity

When the 2-in. cavity first came into use, the outer leaf was connected to the inner by ties of wire or strap steel twisted in the middle to make rainwater drop into the cavity instead of crossing it. However, there was little understanding of the ways in which cavity moisture might affect ties and no research had been done. Some ties were uncoated, some thinly coated with bitumen, and some were galvanized, often rather thinly. The ash mortars which were traditional in some northern parts of Britain proved to be very aggressive on metals embedded in the outer leaf. The process of carbonation, not then well understood (see Section 4.4.2), also took place in mortar (lime:cement mortar in particular), and led to a loss of protection for the steel so that rusting developed. Corrosion was sometimes so severe that the parts of the ties in the outer leaf simply disappeared, leaving no connection whatever. Occasionally the rusting damaged the joint by expansion. This first came to light during inspection of bomb damage in 1940–41, and has latterly been the subject of research at the Building Research Establishment.[1] There is also a British Standard for ties, BS 1243: 1978.

The term 'ties' disguised for a long time the fact that the ties have also to act as struts, and that the stability of cavity walls, even for dwellings, depended on both functions being served. As late as 1946 one notable house-builder told me that in his view ties were only useful while walls were under construction – 'for about a fortnight'. That year I carried out the first research to establish how much strength was needed for ties to function as struts. The studies were unreported, but resulted in the earliest British Standard for ties. This was in the early post-war period.

There have been numerous misfortunes, some serious, when ties have corroded or been used negligently. Sometimes too few have been used or mortar strength has been so scamped that the ties pulled out when subjected to severe wind suction. Some collapses of outer leaf brickwork have occurred.

These and the widened cavities necessary to accommodate greater thicknesses of insulation have focused fresh attention on tie design and specialized types have come into use related to particular forms of insulation. Safe conventions are now well established, while the connectors for large panel systems such as concrete cladding, thin stone, and thin sheet systems are recognized as requiring proper engineering design to ensure that horizontal wind loads get taken back properly to the main structural support system.

Case note

In one remarkable case, two blocks of three-storey flats were built in structurally designed cavity brickwork with precast concrete floors. The block first completed was occupied, but ground-floor tenants complained of a lot of dampness at the base of the outer walls. The second block showed the same fault before occupation and the builder was instructed to open up the inner leaf near any damp patches and clean out the wet debris presumed to be causing them. Two operatives started work on a Thursday, but instead of taking out the usual three or four bricks at intervals for inspection they took out three or four courses all around one wing. Amazingly it all hung together until about 02:00 hours on the Saturday, when a great crash woke tenants in the first block.

Nothing untoward was visible, however, until an early riser, glancing out of his window as daylight came, saw concrete beams projecting through windows in the second block. He went over to take a look and found that most of the interior had collapsed, leaving the outer leaf standing substantially free for its full height.

Subsequent inspection showed that very few ties had been used, just enough to hold it all together for those few hours but not enough to pull down the outer leaf when the inner collapsed. It was very remarkable, but further investigation revealed that the whole building was infested with entertaining stupidities, such as the entry of mains power via a drainage run. The buildings were demolished.

4.2.6 Door and window openings

The inner and outer leaves connect around door, window and other openings in walls and need to behave in a similar manner to each other dimensionally in order to avoid disturbance of finishes. Occasionally this can also be important structurally, for example where the outer leaf is of brickwork and the inner is timber framed. This situation is examined later in Section 4.8.4.

4.2.7 Sound insulation

Sound reduction of nearly 50 dB is typical of brick cavity walls and much the same reduction is given by precast concrete outer cladding with a conventional brick or block inner leaf. It is largely determined by the combined weight of the two leaves because of their firm connection by ties; the windows are then usually the weak points for sound insulation. However, outer leaves of thin stone are lighter than brick, and renderings even lighter. Thin metal or plastic sheet screens offer scarcely any sound insulation and leave the inner leaf to do all the work. The use of open joints for wind pressure equalization reduces the contribution of any outer leaf almost to insignificance. If the inner leaf has to provide most or all of the sound insulation needed, this may be an important factor in selecting the material for it. Ideally, windows and the wall should offer much the same sound reduction if the full value of sound insulation is to be obtained.

4.2.8   Key points for design

- Cavities should always have a gap for ventilation air between the insulation and the outer leaf.
- Stack effect will power it but should be limited to continuous heights of about two storeys. Greater heights give too much power if a cavity fire develops.
- Door and window reveals should always be able to prevent fire breakthrough from the cavity.
- It is almost impossible to keep atmospheric humidity out of insulation and some allowance should be made for the reduced value that results.
- Proper care should be given to the choice of wall ties and their secure fixing.

Sound insulation will at best be what would be expected from the combined weight of the two leaves. Openings in the outer leaf and reduced weight in the inner will be causes of loss.

4.3   Brick and blockwork as the outer leaf

4.3.1   Behaviour and materials

The islands of Britain and Ireland have extensive clay and cement resources and limited supplies of native structural timber, so it is natural that brick and blockwork masonry should have become and remained commonly used materials for walls. The outer leaf is then usually a single thickness of clay or sand-lime (calcium silicate) bricks or of dense or open-textured concrete blocks. Some of these absorb water freely, some moderately and some only slowly. Those materials which are absorbent will often take in and hold quite a lot of rain until drier weather allows evaporation, but the less absorbent they are, the greater will be the amount of free water running down or blown across the face of the wall and available to enter at joints, which it will do by capillarity and/or wind pressure and/or thermal pumping. Thus an outer leaf of absorbent brick or blockwork may wet through only slowly and perhaps only in persistent rain and therefore be relatively protective, while low-absorption, high-strength bricks or blocks may leak rapidly at joints, especially if the mortar is also dense, strong and not very absorbent. Brick/mortar adhesion is seldom good with dense materials (Figure 4.1(a)) and engineering brickwork is particularly prone to this

(a)

(b)

Figure 4.1
(a) Brickwork joints are not watertight.
(b) Disfigurement by alkaline leakage forming calcium carbonate deposits.

shortcoming. The water which emerges from any brickwork can be highly alkaline and will then form deposits of calcium carbonate which are very hard to remove (Figure 4.1(b)). Brick or blockwork outer leaves therefore must always be assumed to be wet and to leak, slowly or quickly and I will sometimes refer to them as the wet leaf.

The outer leaf will also be persistently colder in winter than was the case when cavities were uninsulated and therefore will stay wet or damp for long periods. If the cavity is not adequately ventilated they will also be slower to dry. There will be contributions to dampness from condensate on the inner face or wherever the dew point occurs if that is within the leaf. All this increases the risk of gradual damage by wet-freezing, and the frost sensitivity of outer leaf bricks or blocks should therefore be checked with their maker when selection is made. Weak mortar is also likely to crumble.

Capillaries and gaps exist at joints because of the imperfect bonding of bricks or blocks to mortar. Mortar beds are uneven, they dry and shrink unevenly, dirt can prevent adhesion, or suction in the bricks or blocks may be strong and overdry the contact mortar before it sets. Perpends have particularly poor contact because the buttering is usually poor, first because operatives know that the joint is not load-bearing, and second because no load pressure is exerted on the mortar as it sets, dries and shrinks. If the bricks or blocks themselves have a lot of moisture movement, as some of them do, they will change size a little from wet to dry and hot to cold and this restlessness will gradually cause some further loosening of joints, and insulation in the cavity enhances this activity. The larger the units and the greater their density and the stronger their mortar, the greater their moisture movement and restlessness will be. Then capillarity will operate more easily and wind-driven rain will get through more readily.

The fact that vertical joints do not have wall-weight pressure on them to maintain adhesion during setting argues against the use of stack bond brickwork.

Some measure of air and water leakage has been provided in research by Newman[2] and Brand.[3] Some 90 per cent of air leakage in brick wall systems examined took place through small gaps in joints even though the gaps are typically as little as 0.1 mm in width. Newman noted that as much as 20 per cent of the water falling on a brick wall goes through it by capillarity alone, while the total percentage may get as high as 50 per cent if assisted by wind. Such figures will vary a good deal from one wall to another but they illustrate orders of magnitude that occur. This confirmation of air leakage in brick and blockwork explains why this form of outer cladding for cavity walls has provided some fortuitous ventilation in cavities even after the requirement for air bricks had lapsed.

Thus despite the superficial appearance of being firm and weatherproof, brick and blockwork up to, say, 250 mm or so in thickness is to various degrees fragmented and accessible to water. In a climate subject to intermittent rain there is therefore much to be said for walling materials that can absorb, hold and evaporate rain readily in place of penetration at joints.

Load-bearing brick walls structurally designed for high-rise construction require strong, dense bricks and mortar and are very exposed at the higher levels. Experience shows that they can allow the passage of large amounts of water.

4.3.2 Mortar

When too much water is absorbed from mortar at its interface with bricks or blocks that are laid too dry, the interface contact cement may be left with too little of the water it needs for a proper set and a poor bond will result. A skilled mason can judge quite well the right degree of pre-wetting of the bricks or blocks needed for the job in hand. The amount that some bricks can absorb is surprising; for example, the commons from Oxford clays – the familiar flettons – can apparently absorb individually as much as half a litre.

The ability of mortar to retain its moisture is therefore important. Cement/sand mixes are poor in this respect. The admixture of lime increases moisture retention and is also the traditional and very good plasticizer for mortar. The use of modern aerating plasticizers may mean tidier sites but they do not seem in practice to retain moisture so well and if overmixed they may bubble too much and further weaken the bond.

If neither lime nor a proprietary plasticizer is on-site it is not unknown for operatives to use some detergent. This is not wise. Some contain sodium sulphate and this may introduce a sulphate problem.

Pre-mix mortars are sometimes supplied without their necessary cement. This avoids the risk of pre-setting if the mix gets damp in storage but gives unscrupulous builders an opportunity to undercement the mix. It is hard to detect this by inspecting the mortar when mixed; the mixing itself should be checked. A useful early test is to see if dried and set mortar in joints scrapes away easily with a key or pocket knife after a couple of days or so. If it does, the mortar should be sampled and evaluated quickly. Weak mortar will not hold ties securely and makes a wall dangerous.

The sand for mortar must have a good size gradient. An excessive proportion of fines adds greatly to the surface area needing to be coated by cement for adhesion while too high a proportion of large grains will result in a high ratio of voids needing to be filled by cement. Either extreme thus needs a high cement content, and since it is the cement that shrinks as it gives up its water on drying, high shrinkage and sometimes excessively strong mortar results. On the other hand, if the sand mix needs a lot of cement and it is not provided, weakness results. The ideal is to end up with a mortar reasonably similar in strength and absorption to the bricks or blocks themselves.

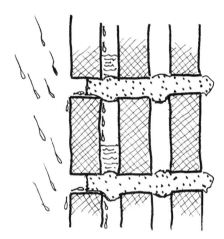

Figure 4.2
Rain on recessed joints with perforated bricks.

4.3.3 Types of joint

The riskiest mortar joint is a recess because water lodges on the ledges ready to enter any capillary or fault, with or without the help of wind (Figure 4.2). A recessed joint is particularly risky with perforated bricks because it leaves too little distance, sometimes none at all, between the mortar face and the holes, and these fill with water. Hundreds or even thousands of these perforations can act as minuscule reservoirs to keep the brickwork damp.

Flush and tooled joints are safer, and there is nothing to choose between them in the present context. Flush joints are simply struck and the face is as absorbent as the mix allows, while tooled 'bucket handle' joints are less absorbent but have the advantage that the mortar is pressed home and therefore seals a little better to the bricks or blocks (Figure 4.3).

4.3.4 Protective modelling of facades

Dry brickwork is not an attainable goal for walls in an oceanic climate but the nearer one gets to it, the lower will be the risk of problems. In

Figure 4.3
Struck and bucket handle joints.

the classical design tradition some protection was given by high modelling in the form of overhanging eaves, cornices, string courses and projecting cills, but the value of this is limited to relatively low-rise buildings, say three or four storeys, and its protective contribution depends a lot on the severity of exposure, a measure of which may be given by the local Driving Rain Index. Where this is low and/or where buildings shelter one another well, much rain falls almost vertically. An eaves projection of as little as 150 or 200 mm at the top of a wall as high as 8 or 10 m can help to keep the face dry and absorbent. In a high Driving Rain Index area, however, it would take a large overhang or eaves to protect even two or three storeys. Therefore on a medium- or high-rise building projections cannot do much good; the upper walls will be subject to the equivalent of a high Driving Rain Index at times and everything will then depend on rigorously good detailing, a well-chosen wall system, good DPC arrangements, and the quality of workmanship.

4.3.5   Expansion and shrinkage

All clay products come dead dry from the kiln and must expand as they acquire their normal in-service moisture content, while concrete products and calcium sulphate (sand-lime) bricks have quite a high moisture content when made and must lose some of it and therefore shrink as they approach in-service equilibrium. For all practical purposes these movements are irreversible in clay and have limited reversibility in concrete, but the sand-lime brickwork retains appreciable reversible moisture movement as Table 4.1 shows.

The realization that clay brickwork expands due to moisture uptake came via Australian research as something of a surprise in the late 1950s. Brickwork had previously been assumed to be thermally but not moisturally active and the expansion which had often bedevilled parapets and long walls in our oceanic climate had been attributed usually to expansion gradually taking place by the accumulation of hard particle dirt in joints opened and closed by cyclic warming and cooling.

Clay products take up airborne moisture relatively rapidly just after manufacture but increasingly slowly with the passage of time, typically taking several years to reach equilibrium. The time scale depends mainly on the hardness of the firing.

The intake is by both absorption and adsorption. The former is the entry of water into pores in the clay, while the latter, which is the slower and more lasting process, is the bonding of water molecules to clay

Table 4.1

Material (in service as external cladding)	Reversible movement (approx. mm/m)	Irreversible movement (approx. mm/m)
Clay brickwork	0.2	(+) 0.2–1.2
Dense concrete brickwork or blockwork	0.2–0.4	(−) 0.2–0.6
Porous blockwork	0.2–0.3	(−) 0.5–0.9
Sand-lime brickwork	1.0–3.5	(−) 1.0–4.0

molecules. The total expansion force is formidable and is exerted both vertically and horizontally, the amounts being normally of the order shown in Table 4.1. Some types of brick or block have more movement than others, which explains the latitude of the figures. The amounts should be obtainable and obtained from the brick- or block-makers; they are not often advertised unless they are very low.

Whether expansion matters or not with clay brickwork in practice depends largely on the circumstances. When it happens to the outer leaf of a cavity wall whose inner leaf does not shrink much on drying, and the scale of operations is that of domestic property in lengths of not more than two or three linked small houses, significant trouble would be unlikely. However, if the inner leaf were of aerated concrete blockwork for example, or timber framing, both of which have quite high drying shrinkage, the differentials are likely to cause problems, and the greater they are, the greater those problems will be. This will typically be evident in differential movement around windows and doors and other connecting positions. Sometimes it has been bad enough to lead to litigation.

With larger-scale buildings different problems occur, depending mainly on how the main frame is built. If it is of reinforced concrete, this has to shrink as it dries and the shrinkage will be about 1 mm/m over periods similar to those for the expansion of brickwork. In practice the conflict between the two movements is resolved by introducing discontinuities in outer leaf brickwork by means of horizontal soft joints, usually at storey height intervals, and soft vertical joints at intervals of the order of 7–10 m. The differentials are reduced when the carcass is a steel frame and therefore not subject to shrinkage, but brick expansion has still to be accommodated both vertically and horizontally (Figure 4.4).

The horizontal soft joints are now commonly managed by carrying each storey of brickwork on a steel shelf, galvanized or preferably of stainless steel, fixed to the main frame, with the filler of the soft joint placed beneath the outward leg. This then accommodates both the brickwork expansion and shrinkage of the main frame (Figure 4.5).

The vertical soft joints need no load-bearing capacity and are simply formed by suitable fillers not damaged by wetting. It is not important to avoid water penetration because the outer leaf will not be water-tight anyway.

It may be asked why the expansion of clay brickwork went unnoticed in traditional construction. The reason is probably the relative softness of traditional lime and lime/cement mortars. These could accommodate some expansion of individual bricks by virtue of mortar softness and its drying shrinkage, while the complex interlocking of the bricks in relatively thick brickwork must also have provided some restraint of expansion.

4.3.6 Outer-leaf fragmentation

The introduction of vertical and horizontal soft joints in outer-leaf construction divides into panels what was formerly continuous brickwork and weakens the wall as a whole. Outer-leaf brickwork formerly acquired stability partly from its continuity, often helped by angled returns of various kinds in plan, and partly from connections to the inner leaf by the ties. Divided into panels, outer-leaf brickwork becomes more dependent for stability on good tie connections to inner-leaf construction, and the inner leaf must then not only be strong and stable in itself but be well secured to the structural frame of the building.

Figure 4.4
Compression cracking in expanding brickwork on a shrinking concrete frame. Note spalling under concentrated compression at DPC line.

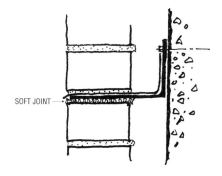

Figure 4.5

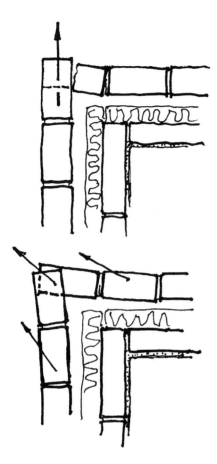

Figure 4.6

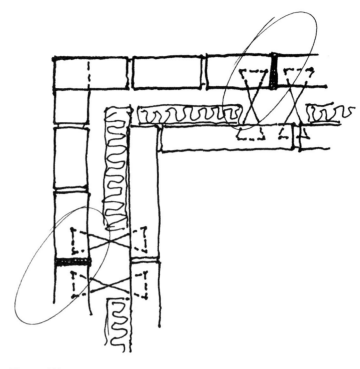

Figure 4.7
Soft vertical joints to avoid corner damage from brick expansion.

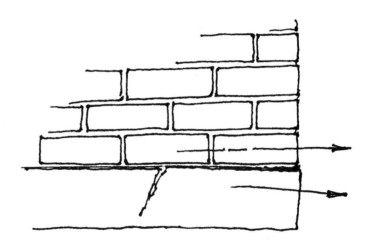

Figure 4.8
Damage to cantilever slab by expanding brickwork on a non-slip DPC.

When horizontal expansion of long lengths of brickwork takes place, corners usually suffer most. The corner moves outwards at an angle, usually using the DPC as a slip plane and forming a sharpened angle. If one of the returns is harder to displace than the other, the one that moves most easily will put a bending moment on the other, typically causing a vertical fracture before continuing its movement (Figure 4.6).

Because of this behaviour it is now established practice to set vertical soft joints near corners but far enough back to allow them to be self-stable. It is also necessary now to place wall ties either side of vertical soft joints and above and below the horizontals for stability at the edges of the brickwork panels (Figure 4.7).

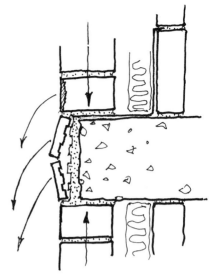

Figure 4.9

Figure 4.10
Brick slips and faces broken away by compression of expanding brickwork.

As regards DPCs acting as slip planes, experience suggests that while bituminous DPC materials have enough in-built lubrication and sheet metals may be smooth enough to allow the slippage, other DPC materials may not do so. If slippage cannot occur, damage of some kind is likely (Figure 4.8).

In the late 1950s and throughout the 1960s the floors of concrete framed buildings were often projected to carry the full width of the outer leaf or, if it was not desired to show the concrete edge, this was finished with thin slip bricks. These created some serious trouble because there were no soft joints in use at that time and the expansion forces in the outer leaf frequently put compressive stress into the slip bricks and caused them to shear off dangerously (Figures 4.9 and 4.10).

In about 1972 the Greater London Council was the first local authority in Britain to require soft joints. A decade or more had therefore passed between recognition of moisture expansion as a brickwork risk and action to deal with it. Much brickwork damage occurred which need not have happened. The industry, collectively, has a high learning inertia.

4.3.7 Calcium silicate (sand/lime) brickwork

Table 4.1 shows that calcium silicate bricks have a much larger reversible moisture movement than clay bricks or concrete blockwork. They also absorb moisture relatively readily and therefore are usually wetted well before being laid so that they will not draw so much setting water from the mortar interface that bond is prejudiced. Without wetting, the bond will be poor.

However, the wetting also means that the wall, when freshly built, will be at least a little enlarged and will shrink somewhat as it dries to a normal moisture content. In prolonged dry weather it will shrink still further but only short lengths will shrink in one piece. Greater lengths will divide up somewhere. A weakish mortar and poor bonding will allow some dispersed shrinkage but at the cost of looser brickwork. If the mortar is strong and well bonded, the shrinkage will be more coherent but there will be a higher risk of cracks.

It is usual to use only a moderately strong mortar – brick makers will usually provide recommendations to go with their bricks – and to place movement joints at positions where one can reasonably expect the shrinkage stress to find weak positions. The best shrinkage relief points therefore are often in the smallish areas between openings one above another in a wall. It is important that the positions of movement joints be well judged to relieve shrinkage; otherwise the stress itself will take charge and cause cracking where it finds weak enough positions. The cracks sometimes follow the brickwork jointing and sometimes just cut vertically through (Figure 4.11).

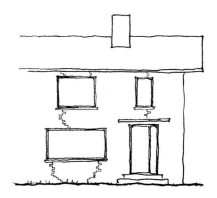

Figure 4.11

4.3.8 Concrete blockwork

Concrete blockwork has been specially developed for outer-leaf construction and is widely used. After its initial drying shrinkage, products of this kind have a residual reversible moisture movement only a little larger but far quicker than clay brickwork, weeks rather than months.

Being larger units than bricks, blocks have correspondingly larger individual moisture movements, but because all products also undergo thermal changes of size, and because the greatest dimensional changes in products which have reversible moisture movements is between the hot/wet and cold/dry states, one should expect from manufacturers and laboratories full hydrothermal movement data for blocks and blockwork, not the data for moisture alone as is usually all that is provided. Again, keep in mind that hydrothermal movements are maximized by insulation in the cavity.

The movement data for an extensive range of products are given in Appendix 1.

4.3.9 Damp proof courses (DPCs)

Early DPC materials, insofar as they were used, were sheet copper or lead, but from the 1920s to the late 1960s bitumen felts usually took their place and were offered with a variety of reinforcing cores – hessian, asbestos, thin sheet lead and latterly fibreglass or polyester. Non-bituminous materials began to come into use after 1945, pitch/polymer in a weight and thickness resembling that of the felts, and various plastics including polythene sheeting.

The conventional DPC cavity 'tray' at the bottom of a wall is formed *in situ*. The DPC material underlies the whole outer leaf and should have a slightly projecting lip. In the cavity it is turned upwards towards a holding tuck-in, usually about 150–200 mm up on the inner leaf. The purpose is to drain cavity water outward. In high-rise cavity walls a DPC tray has usually to be formed at each storey, usually in association with cavity firebreaks.

The location of DPC trays in other positions sometimes needs careful thought and two particular positions need comment. First, where a low-rise part of a building abuts a higher one the outer leaf on the upper wall becomes an indoor wall below the meeting line and a full DPC tray is necessary at the lower roof level, supported across the cavity and with external flashing to prevent water running down and spoiling interior finishes. It may not be an easy type of tray to form if the abutting building has a pitched roof, because the DPC has to be stepped to follow the roof slope, and if the outer leaf is brick-work and the inner is of blocks there will then be three slope factors to be sorted out in the stepping for a buildable, leak-proof tray system – the pitch of the abutting roof, the stepping of the outer leaf and then of the inner.

This problem is not easy to solve buildably. It should always be sketched out 3D and it may need discussion with the contractor as to how best to do it – if indeed it can be done successfully. There have been attempts to devise proprietary stepped trays for this problem but it is not always easy to use them. It is more rational to design the low-rise/high-rise link in a way that makes reliable rain protection easier to do.

DPCs must always be supported across cavities or they droop to form gutters, which usually leak at laps.

It was always regarded as good practice to lay the DPC in the outer leaf with a short projecting lip but because the bitumen in felts sometimes 'bled' a little in the form of unpleasant black runnels, architects in the 1960s began to specify that the edge of the DPC should be set back 10 or 12 mm from the brick face with external pointing in front of it to stop the bleeding, and this was uselessly continued even with non-bleeding DPC materials such as the pitch/polymers.

Those who thought about the bleeding probably assumed that it was due to bitumen being squeezed out by the weight of the wall, but in fact it was usually due to the shearing action that occurs when the wall resting on the DPC moves as expansion or shrinkage takes place. Even when the DPC was set back, the wall movement was usually evident by shearing damage to the pointing mortar.

While bleeding has diminished with newer types of bituminous felts, a largely unrecognized virtue of these materials which the modern bituminous felts fortunately retain is their ability to self-seal under wall pressures and at laps by microflow of the bitumen into the materials between which they are sandwiched. This is not a property shared by the non-bituminous DPC materials. If one has to take brickwork apart to investigate a fault, one soon finds that a clean separation is easier from non-bituminous DPCs than from the bituminous.

If a true seal cannot form there will be capillaries or gaps under the DPCs or at laps, and experience has shown that if the DPC is set back, mortar pointing can then act as a dam behind which cavity water backs up until it finds a way through these gaps or capillaries to defeat the DPC (Figure 4.12). In our numerous investigations of wall problems at BAP there seems to be no record of DPC leakage where bituminous felts

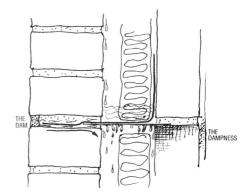

Figure 4.12
Cavity water escape blocked by pointing. Water backs up under the DPC and crosses the DPC support to wet the inner leaf.

were used but many leaks with non-bituminous materials. Some DPCs have been seen set back 35 or 40 mm, which is dreadful.

In a relevant British Standard (BS 5628, Part 3: 1985) the recommendation was made to lay DPCs on a fresh mortar bed, ostensibly to avoid forming capillaries, but experience shows that this is not a very useful or satisfactory practice. The mortar bed is unlikely to be uniform, it is often slightly set before the next course of bricks is laid, and brickwork pressures themselves are not uniformly exerted through the mortar overlying the DPC. Interface contact between DPC and mortar beds is therefore uncertain and capillaries and gaps remain.

Within the cavity itself it has become usual to show the DPC on drawings as a sloping straight line, but this is really the vestigial remnant of a detail which formerly included a sloping mortar bed behind it. Like some other useful traditions this sloping bed apparently seemed insufficiently important to be retained in the face of post-war cost-cutting and it has passed out of use (Figure 4.13).

Without it, however, the slope can never be formed because it is never practicable to draw the DPC up tightly and hold it in the tuck-in on the inner leaf while brickwork is being laid. It always sags into some curved shape (Figure 4.14), and without the sloping mortar bed, laps are difficult to form and seal. As a result they have often been left dry and open and water can and has run through them. Pre-made sloping back-up pieces in plastic or sheet metal have become available for support at laps and a better seal can be made with them.

The distorted DPC frequently collects debris which bridges the cavity. If a mortar bed is laid to receive the DPC under the outer leaf it must also be laid across the cavity, otherwise a shallow trough gets formed into which the DPC sags and then collects water which cannot get out except by leakage at a lap or fault.

A rational alternative to the uncertain slope, one which is better suited to receive the bottom edges of modern cavity insulants, is to form the crossing of the DPC horizontally on the base of the cavity and then turn it up at the back, forming an L-shape against the inner leaf (Figure 4.15).

Greater cavity widths have become necessary as insulation thicknesses increased and the widths of DPCs have had to be increased. Sometimes the material is cut on-site, in the roll, but builders may be tempted to economize by cutting it too narrowly and then find that they have to keep it back from the brick face or omit the tuck-in to the inner leaf. Conversely, they may cut it too wide so that it doesn't fit neatly into the cavity but instead gets squeezed into some sort of S-shape.

Although laps in bitumen felt DPCs self-seal well under brickwork loading, it is better to seal them properly with bitumen. Pitch/polymers need a special adhesive but surprisingly often it has not been on-site or could not be found and was omitted. There is apparently no adhesive at all for polythene except sticky tapes and they seldom seem to be reliably used or long lasting.

Polythene has displayed some other disadvantages. It is flimsy and easily displaced as bricks or blocks are laid, easily blown about in wind and impossible to cut and fold into the shapes necessary for waterproof corners. Frustrated operatives sometimes just bunch it into place. When bunching or welts have been opened up in investigations of leakage there is usually water inside them.

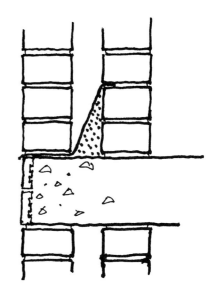

Figure 4.13
Pre-1939 detailing of cavity DPC.

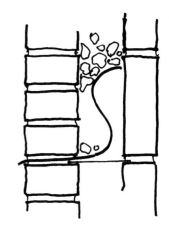

Figure 4.14

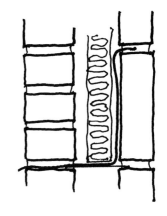

Figure 4.15

Figure 4.16
An asphalt DPC being squeezed out.

Some non-bituminous DPC materials contain a plasticizer. Plasticizers do not combine chemically within a product but remain volatile. They may then be absorbed by other building materials such as mortar and concrete or they may migrate away from points of high pressure to those of low or none, e.g. within the material in the cavity.

For various reasons brickwork does not exert uniform pressure on its bed, and, of course there is no pressure on a DPC in a cavity. In some instances it appears that plasticizer has moved within DPCs from brickwork to cavity, causing depletion in the brickwork pressure area and enrichment in the cavity. Depletion reduces volume and could therefore be the cause of transverse shrinkage cracks sometimes found under the brickwork. Longer, longitudinal splits have appeared in cavities and are perhaps due to a weakness caused by over-enrichment. With wider recognition of the problems, improvements have been made in these materials.

Asphalt has sometimes been used as a DPC material but for reasons discussed in Chapter 9 some asphalt mixes can flow more easily than others and are unsatisfactory as DPCs (Figure 4.16).

DPC corners

As mentioned in the discussion of basements, there is no geometrically leak-proof way of turning a DPC tray around either an inward or an outward corner, and therefore there is no 'right' way for operatives to do them. There will always be a weak point where the three planes meet, so what is needed for corners is a pre-formed shape, feather-edged to avoid unfilled gaps caused by the thickness of the corner material and with a radiused turn-up between horizontal and vertical faces to receive the felt, because this cannot be sharply bent in cold weather without cracking. If pre-formed corners have sharp corners and are used with radiused felt there must be leaks, and some have been seen through which one could put a finger (Figure 4.17).

Failing the use of pre-formed corners, an inner turn can sometimes be made water-tight by generous use of bitumen at the vulnerable point, but it cannot be very reliable.

On several investigations attempts have been seen to protect corners by using a proprietary material in the form of a flexible sheet plastic with an adhesive on one face. It can be folded and stuck in place and initially can often be made to look water-tight. However, plastics have memories, and if they have one of being flat, before long they try, usually with some success, to become more or less flat again, which breaks the adhesion so that leakage takes place.

When the run of an inner leaf is interrupted by the intrusion of a column or cross-wall, proper provision must be made for the continuity of the DPC tray and its tuck-in around the intrusion. It must not be interrupted.

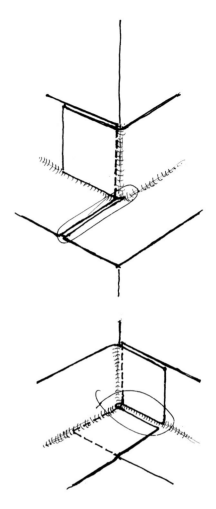

Case note

A building inspector unwisely required concrete party walls to intrude and cross the cavity so that the outer leaf could be tied directly to them. No detail had been worked out for routeing the DPC around these interruptions and the operatives simply tore the tray where it met the column.

The latter had been given a vertical DPC on the contact face with the outer leaf but this was split to accommodate the ties. Water on the tray itself went inside at the torn positions and wetted the internal corners of rooms. Water running down the vertical DPC strips on the noses of the cross-walls got into meagrely mortared vertical joints in the outer leaf and found its way inwards or outwards further down, depending on the vicissitudes of the horizontal mortar beds. Both leaves of all the outer walls had to be stripped out and rebuilt at substantial cost.

All such DPC corners and meetings of vertical and horizontal DPCs should be sketched 3D to ensure that they are buildable.

Figure 4.17

4.3.10   The junctions of DPCs and DPMs on slabs

In the preceding chapter the damp-proofing of ground slabs for low-rise construction was discussed. The necessary slab DPM must meet and be joined to the wall DPC in a water-tight manner.

Two situations typically occur. Consider first the situation when the ground slab is to go through the envelope wall to its outer face. The DPM on the central areas of the slab should not be laid before the walls are built because of the risk of damage to it during their construction. Instead a DPM should be laid under full width of the outer walls with a selvage projecting inward to which the main DPM can later be bonded (Figure 4.18). The projecting selvage should be bonded down to the slab to minimize risks of damage during construction. The DPC and the DPM must be of the same material for reliable bonding, and the upstand to complete a DPC tray would be added when inner-leaf bricklaying begins.

If the slab stops at the inner face of the envelope wall to avoid cold bridging, much the same policy can be followed, but one must keep in mind that on clay soils the foundation under the outer wall is likely to settle a little more than the slab, sometimes by as much as 8 or 10 mm,

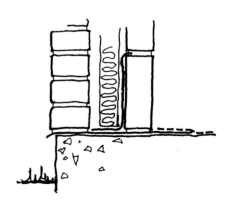

Figure 4.18

and no risk should be taken of having the DPM shear along this line. Avoid bonding tightly across it therefore.

DPCs for window and door openings are discussed in Chapter 8.

4.4 Pre-cast reinforced concrete cladding

Concrete cladding panels were first used in the last quarter of the nineteeth century when W. H. Lascelles developed a house-building system of concrete panels on timber frames, and Norman Shaw used concrete units on a house in Croydon. After the First World War several systems were tried, one on the outskirts of Cambridge, towards Milton, and another in Welwyn Garden City and others in the south-west. After 1945 there was a more determined effort to establish their use for housing but it eventually lost steam.

All this early work suffered problems with reinforcement, casting, and the use of chloride accelerators for setting. Faults have also been found with fixings pre-set into the pre-castings but not precisely matching the tie locations on the main building fabric. The subsequent improvisations have proved precarious.

4.4.1 Quality and life expectancy of reinforced concrete

In a paper read to the RIBA Annual Conference in 1953 by myself and E. D. Mills,[4] we drew attention to the fact that reinforced concrete on some European buildings had spalled badly in as little as 15–18 years, even in sheltered positions, and that it could be unwise to rely upon it for a trouble-free life of more than perhaps 25 years. Subsequent experience in the climate of these islands has demonstrated that its life has varied widely; concrete railway sleepers take a lot of punishment but are very durable while some pre-cast cladding has come apart dangerously in less than 10 years. Considering the cost and nuisance of premature cladding replacement and the fact that its degradation may go unnoticed until some falls off lethally, the only rational policy is to design, specify and manufacture for a life expectancy of the order expected for the building itself, normally at least 50 years. This needs care all the way through design to manufacture, delivery and fixing.

The usual agent of panel destruction has been the corrosion of the reinforcing metal when it has been mild steel. Where mild steel continues to be used for reinforcement, the protection necessary for a long life is high-quality, fault-free concrete with consistently adequate thickness of cover on the fronts and backs and around the edges of the pre-castings. This asks a lot, and a better option is to use galvanized mild steel or preferably austenitic stainless steel.

Reinforced concrete design is one of the main branches of structural engineering and the literature of that profession is rich in guidance and research on concrete quality. It is neither appropriate nor practicable to review it in detail here but common ground is necessary for rational discussion by members of the design team and accordingly the main elements are outlined below.

4.4.2   Quality factors

The aim must be a product with low permeability and low drying shrinkage. Hard, low porosity aggregates must be used – stones vary in hardness and permeability – and the size gradient of the aggregates, and the amounts at various sizes, should be such that if they were tumbled together dry, the voids would be relatively few and small, like a macro-version of sands for good mortars. The ideal, never attainable, would be a fit so good that only a thin coating of cement paste on all the aggregates would bind the whole into a good approximation to solid stone. Both shrinkage and permeability would then be minimal because it is the cement paste that mainly causes shrinkage as it releases its excess water on drying, and it is the cement paste that can become permeable in the course of losing this water.

The proportion of the finest aggregates is very important. If the amount in the mix is too small, voids will increase and require more cement paste to fill them, while if the amount of fines is needlessly large the total surface area of aggregate needing to be coated will increase disproportionately. Either way there will be a need for a higher proportion of paste and this in turn will cause more shrinkage. On the other hand, if not enough paste has been incorporated to cope with these needs, voids will be poorly filled and the coating will be meagre, either of which will weaken the concrete and also make it more permeable to moisture.

Case note

A large programme of long, narrow pre-castings had to meet specified strength tests. A percentage of the units mysteriously bent like bananas under the test loads. A short, intensive study revealed that these units were all made on Friday afternoons, which was when the aggregate bins were cleaned out and a large residue of fines came down into the mix. Apparently there was then not enough cement to coat all the aggregate properly and fill the voids.

If high shrinkage is created, a specific risk is the opening of micro-cracks, for example around exposed coarse aggregate. These not only reduce the effective cover for steel because of the deep access they provide for the water but they also enable surface aggregates to be more easily dislodged by frost.

Plasticity

The plasticity of the mix is another of the balances to be struck for good-quality pre-castings. When concrete is vibrated in the moulds it must be plastic enough to enclose the steel intimately but not so sloppy as to let the coarse aggregate settle out. The distribution of aggregates must remain reasonably consistent.

Case note

Wall components of room length were designed to form a rather thin sloping cill along their upper edges (Figure 4.19) and the moulds were therefore doubled to provide an inner and outer profile. The cills developed cracks and investigation

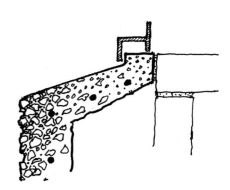

Figure 4.19

showed that the coarse aggregate had settled out of the cills during vibration leaving an excessive proportion of fines in them and therefore a predisposition to shrink. Since the cills were also too thin, the steel in them was put at risk from insufficient cover as well.

The amount of paste and the amount of water in it – its water/cement ratio – are important factors in determining the degree of plasticity of the mix, but the shape of the aggregates is also relevant, the roundness of gravel being more conducive than the angularity of crushed stone. Unfortunately the readily accessible gravel resources of Britain are being used up and this is increasing overdependence on crushed stone. The porosity of stone can also be a factor because it can take up water intended for the paste and thus affect the water/cement ratio.

Carbonation

The references to the importance of the cement paste bring us to the vital factor in the long-term protection of steel, the rate at which carbonation is able to develop in the concrete. The process itself requires more than an outline explanation in order to make its importance clear, together with the crucial role played in its development by the water/cement ratio of the paste.

The two main cementing constituents of ordinary Portland cement are dicalcium silicate and tricalcium silicate, and when water is introduced to make the paste, they react with it to form calcium hydroxide. The reaction is known as hydration because of the water involvement. Calcium hydroxide is strongly alkaline – it has a pH value of about 13 – and the alkaline environment it provides for the steel prevents rusting by the quick formation of a protective film of iron oxide.

The preservation of this film is essential for a trouble-free life for mild steel, but a destructive agent is at hand to undermine it. Carbon dioxide is one of the common gases in air (see page 13), and when it makes contact with calcium hydroxide another reaction takes place in which the product is calcium carbonate. This is the specific process known as carbonation and its danger for the reinforcement is that it neutralizes the alkalinity of the affected paste and thereby destroys the protection it gave to the steel. Therefore carbonation must not be allowed to move rapidly inwards.

The atmospheric carbon dioxide gets into the concrete by permeation, for it is a gas exerting normal gaseous pressure and probably the thermal cycling of the pore structure near the concrete surface as its temperature changes helps gases to enter by a micropumping action. Low-permeability paste is therefore vital for durable protection of the steel. Although the porosity of stone aggregates usually has only a limited effect on the overall permeability of concrete, low permeability is a good attribute for them too, as noted earlier. In a few cases high-permeability aggregate has contributed critically to early steel corrosion.

However, the key factor is the water/cement ratio of the paste because if its water content is too high, the process of drying makes the hardened paste too shrinkable and porous, resulting in generally easier access for carbon dioxide and other gases.

The other relevant gases are water vapour and atmospheric oxygen, and together with CO_2 they begin to undermine the integrity of the concrete. The CO_2 and the calcium hydroxide combine to form the carbonation front which gradually progresses inwards towards the steel, reducing the pH value of the alkalinity as it penetrates while at the same time the CO_2 is also dissolving into the water vapour to form carbonic acid. This is a weak electrolyte and when carbonation has eventually lowered the protective alkalinity around the steel, electrolytic currents can flow through the moisture from anodic to cathodic sites on the metal. The presence of the atmospheric oxygen then facilitates a cathodic reaction to form hydrated iron oxide, which is rust. The pH values may by then have dropped as low as 10 or even 9 around the steel, and these are low values.

The rust occupies a greater volume than the metal and the powerful force of its expansion has a bursting effect on the enclosing concrete. Initially it may show itself simply as cracks in the panel following roughly the lines of the reinforcement, but these then facilitate the entry of moisture, which is followed by staining and the possible detachment of shells of concrete. If the rusting is initially localized these may be small, but as the attack becomes more general the detachments become larger, weighing as much as 50 or 75 kg, and potentially lethal. Detachments are most likely in freezing wet conditions when ice can form behind them, or in hot sunshine, when expansion can loosen them.

Carbonation takes place on both sides of castings and, perhaps surprisingly, it usually moves inwards quicker on the sheltered side than on the weather face because it is encouraged by warmth and because it is easier for CO_2 to enter a dry face than a wet one. The process likes RH values between 55 per cent and 75 per cent, which is why cover must be equally good on the backs and fronts and around the edges of castings.

It is not possible by specification to relate closely the quality of concrete to rates of carbonation and therefore one cannot specify exact amounts of cover to give particular life expectancies. Penetrations of 35 or 40 mm have been known to take place in a very few years in poor concrete. What can be said reliably is that it is desirable to reduce water/cement ratios to below 55 per cent, even 50 or 45 per cent, for true depths of cover of, say, 40 mm, using good types and mixes of aggregates and good Portland cement. By 'true' depths I mean the depths after any allowance for surface microcracking.

Chlorides

The presence of chloride ions in reinforced concrete increases the aggressiveness of rusting. There are always very small quantities of chlorides in water and aggregates and this is accepted as an unavoidable norm, but larger amounts can quicken the setting of concrete, and proprietary admixtures containing calcium chloride in particular have frequently been used in the past in the manufacture of pre-castings to reduce the de-moulding time. It speeds up the factory throughput and thereby reduces costs; it can also speed up failure.

The practice should be behind us now for the risks are universally recognized and additives containing chlorides are 'out' for any type of reinforced concrete. However, it has been noticed that when pieces of a pre-casting get knocked off during factory or site handling – as may happen to edges and corners particularly – it remains a temptation to do repairs with concrete containing rapid hardening additives. The advantage to the manufacturer is obvious, but chloride ions migrate readily in the amounts of moisture normally present in concrete and a repair will put them closer to steel. Chlorides should never be allowed, even in repair mixes.

Chlorides attack concrete in the following manner. They break down in moisture to form a very strong electrolyte which further reduces the alkalinity of the concrete, increases the electrolytic flow of corrosion currents and creates the enhanced rusting.

Case note

On a big run of cladding units some of the stone aggregate was large and porous, permeable to both air and water. It was sufficiently large to be exposed sometimes on the surface and also to be in contact with reinforcement. The concrete had been made with additives containing calcium chloride. In a very few years large spalls began to fall and all the buildings concerned had to be overclad.

Chlorides are salts. Seaside environments are salty and poorly made pre-castings are therefore likely to have an even lower life expectancy near the sea. Aggregates from marine environments can also be damaging if not thoroughly washed.

Faults

Local penetrations of what may otherwise be good protection must be avoided because localized rusting can develop and spread. Weak spots may occur, for example around spacers used to keep reinforcement away from the mould face, or where cramps or bracket fixings or lifting sockets are cast in, or where a large piece of aggregate goes from the face to the steel, and they are particularly likely at repairs because it is difficult to bond these 100 per cent. The risk of damage to edges and corners has been mentioned, and a lot can be said therefore in favour of rigorous rejection of any but the most minor repairs, either before pre-castings leave the factory or on arrival at site or after fixing in position. This needs careful inspection by someone who knows what to look for. There will be an increase in rejection rates if this is done, but it is what is needed for safe cladding.

Case note

In one case an engineer-inspector in the factory noted that every pre-casting in a large programme had faults and he therefore held them back. Unfortunately the architect underrated their importance and criticized the engineer for being unduly fussy and thereby delaying the programme. The engineer fell silent. The castings were accepted on-site in the knowledge that these faults existed, and were put in place by a contractor who was also aware of them. Trouble naturally developed and repairs had to be made in situ, but it was impossible to give any assurance that the claddings would then have anything better than a short life. See also Section 4.4.4.

If damage occurs but the breaks are clean and total so that they allow a tight refit, a repair may be feasible without fresh mortar. A simple and good technique then is simply to apply a watery cement grout to both faces and keep them pressed well together long enough for a set to take place. The grout must be really watery, not creamy.

Figure 4.20

Steel positioning

It is one thing to specify the amount of cover and quite another to ensure that the steel is where it is intended to be. The positioning is sometimes done very carelessly.

Case note

A case occurred where pre-castings were to have grooves about 20 mm deep making a pattern to match shaped grooves at joints. Steel was to be bent inwards around these grooves to maintain cover (Figure 4.20), but it seemed possible that this inconvenience might tempt the manufacturer to take a chance and run it straight past. Cover meter readings on delivered panels confirmed that this might be happening, so a casting was broken open and confirmed the suspicion. Cover was down sometimes to a few millimetres at grooves, the very position where shrinkage stress could open cracks. Some of the castings were to be load-bearing and were already in place and had to be removed. A large number of others awaiting delivery were ordered to be destroyed.

Steel positioning should be systematically inspected during cladding manufacture and a small percentage of castings should be broken open for checking before shipment to site in order to be sure that manufacture is consistent and satisfactory. Look particularly for the following:

- That filling of moulds has not displaced the reinforcement
- That vibration for consolidation has not shifted it
- That no steel is getting less than the intended cover
- That voids are not being left around the steel
- That vibration is not segregating the coarse aggregate
- That microcracking is not occurring. This needs particularly careful examination.

The break-up of a casting for inspection needs to be comprehensive to check matters like these. Cover meters can tell one about depth of cover but not about the other details.

4.4.3   Ribbed surface treatments

Two other types of faulting have been encountered during investigations, both on ribbed finishes. Flat panels were often found to dirty disagreeably in Britain's oceanic/industrial atmosphere, and strongly defined ribbing became popular for a time as a way of developing a masking pattern of light and shade on them. The ribs projected 50–75 mm with perhaps 40–50 mm separation at the base.

Case notes

One form of ribbing was textured on the outer edges of the ribs by sideways hammering while the concrete was fresh and green from the moulds. Not surprisingly this caused incipient cracking inwards along the angle between the sides of the ribs and the base concrete.

A second form used selected coloured aggregates exposed on the outer edges of the ribs. The stone was dropped into the rib recesses in the moulds before concrete was poured, but even with vibration the cement paste often could not work well down into it and some of it remained loose enough to be dislodged in freezing, wet weather. This can be dangerous.

Apart from the damage that hammering can cause, it is not good practice to form a sharp angle at the base of the ribs. It causes shrinkage stress to concentrate along the corner which opens up cracks that propagate inwards as the concrete matures and dries, giving easy penetration for water. Rounded meeting lines reduce the risk (Figure 4.21).

This applies to all inner corners on concrete castings and references will be made later to other circumstances where this behaviour occurs, notably in sharp corners of door and window openings. It also takes place in other materials such as plastics and metals where the stress may be introduced by shrinkage, thermal cycling or vibration. It is a chief reason for rounded corners in windows on aircraft, for example.

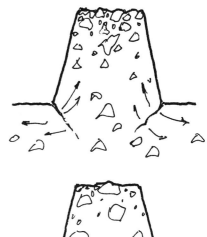

Figure 4.21

4.4.4 Tiling and mosaic on pre-castings

Tiling has been used on pre-cast concrete cladding, sometimes with unexpected results.

Case notes

White undercut tiles were laid face-down in the moulds and the concrete was poured to a 100 mm thickness directly on them. After a few months in place the panels developed an outward bow. The evidence eventually indicated that because evaporation was restricted through the tiling, the preferential drying of the back face was causing it to shrink quicker than the front, creating a permanent set. Moisture expansion of the tiling will also have played a part by expanding the front face. Different treatment of the backs and fronts of castings by finishes that cause different drying rates or which themselves have different behaviour from the concrete is unwise.

A second case concerned two-storey claddings. They were heavy and needed a lot of reinforcement, so much in fact that in many places only small amounts of concrete got through it to grip the tiles during casting. The poorly held tiles soon revealed themselves by coming off as the drying shrinkage of the concrete took place and the tiles expanded, and this in turn revealed exposed or poorly covered rusting steel behind. Repairs were carried out but it was realized that the claddings and the tiles would now both have abbreviated lives.

Figure 4.22

Mosaic, as used commercially, is usually pre-fixed to paper sheets with a water-soluble adhesive. When it is to be the facing for pre-cast concrete, the mosaic sheets are laid in the mould paper-side down and the concrete is poured directly on to them. The paper is subsequently removed, a grout filling for the joints is applied by squeegee, and any excess is wiped off.

Mosaic is made with chips of stone, glass or hard-burnt clay and neither the pieces themselves nor the grout has any significant tendency to absorb moisture and expand like clay tiles, but it can inhibit drying on its side of a cladding unit and create bowing by preferential shrinkage of the other as in the tiling case mentioned previously. Neither tiling nor mosaic can be assumed to contribute to the protective cover of reinforcement, as a case illustrates:

Case note

An assumption had been made that mosaic was effectively waterproof and that the reinforcement therefore did not need much additional cover. Mosaic is not waterproof and the steel soon began to rust, causing spalls of concrete to fall (Figure 4.22). See also the discussion of mosaic in applied finishes, Section 5.4.3.

4.4.5   Stone finishes on pre-castings

When concrete pre-castings are to be manufactured with a stone finish the selected stone slabs, usually 30–50 mm thick, will normally be laid in the mould first, with the joints taped to prevent concrete grout getting into them and with the back faces of the stone given a bond-prevention coating to avoid stress created by differentials in concrete and stone behaviour.

The fixings for the stone are usually stainless steel dowels set at opposing angles to ensure a positive lock. The dowels will probably have a neoprene setting in the stone as a further allowance for differential movements. It is important in placing the dowels in the concrete to ensure that their location does not prejudice the cover for the reinforcing steel; the dowels have to reach 50 mm or more into the concrete.

As with other applied finishes, stone should not be assumed to contribute to the protective cover for the reinforcement.

Artificial stone

Castings can be given artificial stone finishes either as whole 'stones' or as a surfacing material for body concrete. Either way the material requires very skilful manufacture and the industry concerned has not got a trouble-free record; but see also Section 4.6.2.

Much of the skill in forming a successful artificial stone finish directly on the body concrete is in getting the composition of the two products compatible, together with careful curing. As both are in fact concrete, in principle both should be able to behave similarly if properly specified.

It would be advisable to see some of a firm's products which have been used for several years on buildings before placing a contract.

4.4.6 Brickwork facings on pre-castings

There is no absolute reason why brickwork facings should not be used on pre-castings, with the pre-formed brickwork placed in the mould first and the concrete cast directly on it, and some very complex work of this kind has been done. However, it is essential to limit the water absorption of the bricks to a maximum of 7 or 8 per cent and to use a slightly soft mortar; expanding brickwork on shrinking concrete is a recipe for trouble. The trade recommends cutting dovetailed grooves in the backs of the bricks for a mechanical key and this reflects the element of perceived risk.

4.4.7 Fixings, joints, size, and shape of cladding panels

Fixings, joints, size, and shape are largely interdependent factors in concrete component design but their interdependency is complex and can best be understood through individual discussion of them.

Fixings

The dead loads of cladding units have to be supported by the carcass of the building, and the positive and negative wind loads must be transferred to it. Anchorage design must therefore be done to suitable safety factors.

The dead loads are usually carried either by casting flanges on the backs of the units designed to rest on beams of the frame or by stainless steel angles reaching under the bottom edges of the units or engaging a step cast on their backs. The horizontal forces due to wind are transferred to the frame mainly by restraining cleats, sometimes four of them doing the whole job, sometimes two or three in association with the load-support fixings. The reinforcement of the units has to be designed for the dead loads to be carried, the loads to be transmitted by the cleats, the handling stresses, and the wind forces. These are structural matters for engineering design.

Fixings must have some three-way adjustability because the cladding must line up as intended but the carcass framing of medium- and high-rise construction cannot reasonably be expected to be accurate. Verticals will want to wander at least 10–15 mm, plus or minus, per storey, and horizontals will not be much better. Steel framing is generally more accurate than *in-situ* reinforced concrete but the use of concrete casing for fire protection of steel can introduce imprecision, and steelwork itself is not necessarily as accurate as might be supposed.

Reinforced concrete has the added characteristics that the verticals will shrink over their first five or six years and the horizontals will sag plastically. As remarked earlier, the vertical shrinkage is likely to be of the order of 1 mm/m and the plastic deflection of horizontal beams and slabs will be as much as 10–15 mm in a span of 5–7 m, depending on the dead and live loads to be carried, the concrete and reinforcement specification, and the shape of the structural element. The three-way adjustability therefore needs to be of the right order to accommodate the discrepancies to be expected, usually 25–30 mm.

When support is by a flange on the backs of pre-castings, the fixing usually gives two-way adjustability horizontally. Vertical adjustment has usually been done by shims, with dry-pack mortar rammed into the gap created by these to take the main weight. In practice it has often not been as easy as it sounds; the units are heavy and have to be supported sensitively while positioning is done and fixings tightened, and dry-pack cannot be rammed hard if there is no stop against which to ram it or if the gap is too narrow to get it and the ram in. A further disadvantage of flange supports on castings is that if the cladding joints allow rain to get in, it may lodge on the flange and run through at the bolt positions where it is then inside the building's wetting defences.

When steel angles are used for support, they are fixed to the sides of the main frame, and bolting systems giving three-way adjustability are normally used. It should not be difficult to place the angles accurately, and if the castings are also accurate, final adjustments ought not to be a problem.

If continuous runs of steel angles are used, a DPC strip can provide a nicely cushioned seating and will cover usefully the end-to-end joints in the steel.

A problem when cladding a reinforced concrete frame is that the deflection of support beams over the first years can cause panel units to tilt unevenly and this risks some crushing at joints. The vertical shrinkage of concrete framing should be of no consequence with concrete cladding since this will also be shrinking. Load-bearing pre-cast cladding has not been used often and there have been some problems with it. One system seen used wall-size units, solid for party walls and with an insulation core for externals.

Case note

Each casting incorporated a pair of 25 mm threaded pins in the top edge, intended to enter sockets in the bottom edge of the next casting above, using nuts for levelling. The pins passed through perforations in the edge of the floor, and dry-pack was inserted between the floor and the bottom edge of each casting to share its load with the pins (Figure 4.23).

Two problems occurred. Sometimes the pins and sockets did not line up but the pins were too stiff to adjust, so various astonishing fudges were used to surmount the difficulty: some pins were bent over and others were burnt or sawn off. No corrections were made.

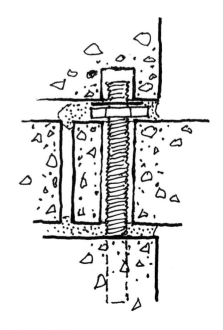

Figure 4.23

In the outer wall units the insulation caused cyclic thermal bending of the castings as the outer face changed temperature and at the time of the investigation it was apparent that this had been grinding away the dry-pack over several years. Not at all satisfactory, and much repair was needed.

When openings for doors and windows are made in pre-castings they are, of course, usually formed with right-angled corners. It has already been noted that sharp angles concentrate shrinkage stress and unless this is countered in some way it will typically open a crack which runs away from the corner at about 45° and propagate for 150 or 200 mm or more as thermal cycling takes place. A customary way to prevent this is to use reinforcement which crosses the line of propagation. Ideally this could be associated with rounding the corners of openings but this is not often practicable (Figures 4.24 and 4.25).

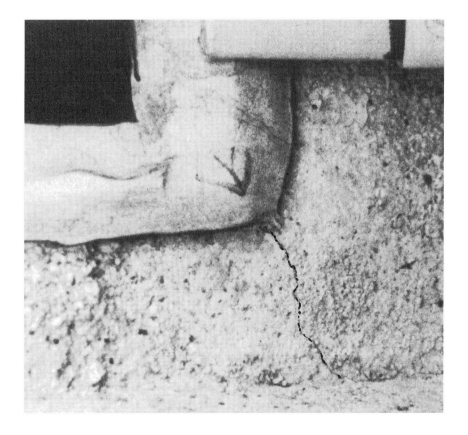

Figure 4.24

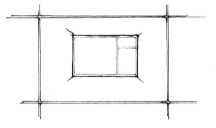

Figure 4.25

Joints and unit size tolerances

Several concepts for closing off joints were introduced in the 1960s and early 1970s, wind-break slips, tubular inserts, drainage grooving of component edges, and others. Few were successful; most survived only a short time and rectification was difficult or impossible. Wind-break slips broke up and could not be replaced. Mortar jointing cannot be used where concrete component dimensions exceed a metre or so because of subsequent thermo-moisture movement of the castings and mastic sealants have become customary, usually of silicone.

A sensible joint width at which to aim is usually 10 mm unless the units are very long. However, with pre-cast cladding one does not specify assembly by joint width because cladding units cannot be shuffled about for fixing. Each unit must have an allocated space and the joint gaps are the spaces left between them.

Thermo-moisture movement during the life of the cladding will be of the order of 0.5 mm/m, plus or minus, and this has to be taken into account in determining the overall size of units, the casting tolerances for edge straightness, and unit squareness. Edges must not be so wavy that movement can overcompress the sealant or cause a damaging solid contact when components expand nor should they be able to overstretch the mastic anywhere when they contract (Figure 4.26). Therefore:

Figure 4.26
An appalling example of pre-casting which created much ugliness and numerous problems.

- Largest unit sizes should probably be limited to about 2.5 m.
- Overall lengths and widths should be accurate to 1.5 mm plus or minus, measured at the corners of moulds or components because these are the only definable positions for the purpose.
- For straightness, edges should not deviate more than about 1.5 mm plus or minus from the overall sizes determined at corners.
- Corner-to-corner diagonals should not differ by more than about 2 mm for squareness.

Mastic has to be applied against a resistance so that sideways pressure develops to make contact/adhesion to the concrete. The resistance is normally provided by pre-formed ethylene foam strips squeezed into place beforehand. This has to be done fairly accurately because it determines the depth of the mastic and this is important because the shape of the mastic in section should end up approximately twice as wide as it is deep, for this ratio stands up best to the cyclic compression and stretching that takes place at joints (Figure 4.27).

Accurate setting of the ethylene foam may be helped by providing a small shoulder, no more than a millimetre or so, on the component edges as shown in the figure, but this will lose 2 mm of gap width between the concrete faces unless the nominal exposed width is increased accordingly.

Mastic seals will need replacement at intervals depending on their rate of degradation. Twenty to twenty-five years would be a reasonable life on present expectations for good mastics, so the facade should be designed to be accessible for maintenance.

All this emphasizes the fineness of balance to be struck in joint design, accuracy of manufacture and the skill with which mastic must be applied. Even when mastic application is well done, however, its adhesion is never 100 per cent; concrete has a slightly porous surface, it will usually have dust or dirt on it, and application pressure will be uneven. True adhesion of the order of 60 per cent or 70 per cent is probably as much as one can expect. Some leakage will occur and gradually increase over the years, but if cavity design has been sensible, and there is ventilation, this should be of little or no consequence.

There should be discussion with the jointing sub-contractor before the joint design is finalized for component manufacture.

Figure 4.27

4.4.8 Moulding and casting

Moulds

It must be assumed that initial component shrinkage on drying will be of the order of 1 mm/m and allowance for this should be made in mould sizing for component dimensions upwards of a metre or so. If sizes and tolerances for joints are to be achieved within the limits described, the moulds must be very strong to stand up to the stresses generated by vibration of the concrete over the intended production run. Sheet steel on steel frames has not generally met these needs and the use of robust timber with a low moisture movement seems to have proved more reliable.

All moulds deteriorate in use and should be checked periodically during a production run – go/no-go jigs are useful for the purpose – and there should be an agreed basis for deciding when to discontinue the use of moulds.

It is sensible to taper moulds slightly at the edges and to radius corners slightly to avoid de-moulding damage. If de-moulding oil is allowed, it must be of a type that will not stain the concrete.

When components are being manufactured the full moulds may be trolleyed from the vibration table to a curing area. Sometimes it causes back-and-forth oscillation of the concrete to take place when its initial set is beginning, and this leads to permanent thickening at the ends of components. Dimensions here may need checking, especially if the profiles are critical for mating to other components. Again a convenient way of checking is to use a go/no-go profile.

Curing time and transport

It can be a temptation to use components too soon after de-moulding in order to reduce storage costs. A time gap should be specified which is appropriate to the sizes and thicknesses of the components – preferably at least two weeks – so that risks of edge-damage and microcracking by handling are reduced and some of the initial shrinkage can take place.

Where manufacture is done off-site, there should be specifications for safe transport. Generally components should be transported on edge to avoid the cracking risks of flat delivery, although panel units are sometimes designed with rib-strengthening on the back to reduce this risk.

4.4.9 Parapets

Parapets are mainly discussed in Chapter 9 on flat roofs, but one point needs attention here. Parapets have sometimes been designed as part of a pre-cast cladding system simply by projecting the top run of units above roof level, where they are then exposed on both sides. This is not a reliable arrangement for at least four reasons. First, the mastic jointing has to be carried over the top and the exposure for it is too severe, abbreviating its life so that leaks occur in a vulnerable position. Second, the roof membrane skirting probably has to be tucked into a groove along the backs of the panels but if the skirting material is asphalt or a roofing felt the movement of the concrete, enhanced by its severe thermal exposure, can soon fatigue and split the skirting material at the joints, allowing water to get in under the roofing. Third, cracks sometimes develop in the concrete which run down behind the groove and let rainwater bypass the skirting. Lastly, the mastic in the joints on the backs of the panels cannot be sealed reliably to the top of a bituminous skirting where it crosses the gap, and little openings develop. It is much better to go no higher than a low upstand and to form a coping in a way that can protect joints and any cracks that may develop in it. (See also Section 9.3).

Generally, the horizontal joints should accommodate a normal DPC while the verticals get mastic.

4.4.10 Total prefabrication

Almost inevitably, total prefabrication became one of the ideological and industrial goals in the housing reconstruction boom of the 1960s. It has not survived, but some fundamental difficulties that became evident are instructive long term.

The concept typically relied on load-bearing pre-cast concrete cross-walls, pre-cast floors (usually in wide panels) and pre-cast outer wall units of room wall size, with core insulation. Floors spanned between cross-walls and their reinforcement was knitted to that of the cross-walls with a grout infill poured around the knitting. There was a levelling arrangement and usually dry-pack. The outer wall units were similarly knitted in to the ends of the cross-walls and locked in by pours of grout, with similar linking of wall units to the edges of floors, creating a monolith.

It is astonishing in retrospect to contemplate the optimism that could ignore the shrinkage of concrete and subsequent reversible component movements, the high shrinkage of grout fillings, and the thermal bending of wall castings with insulation cores. They all caused leakage.

Case note

In one investigation a water spray on a wall at an upper floor level produced generous inward leakage along cross-walls, floor joints, and floor-to-wall junctions two storeys below within half an hour. As a further misfortune, stress cracks developed at 90° window corners allowing rain and vapour to enter the core voids containing insulation. The corner cracks went through to the inner face and caused staining. It was all very difficult to deal with.

4.4.11 Design liabilities

Good engineer–architect collaboration is clearly vital in the design of pre-cast concrete cladding. The cladding is part and parcel of any architectural concept for a building on which it is to be used but its engineering design is fundamental to its structural success and durability, and between the two disciplines buildability must be ensured for good on-site work and freedom from subsequent misbehaviour.

Whether a formal contract spelling out the division of liability between architect and engineer is needed is a personal question for those concerned, but it should be kept in mind that if something does go wrong and litigation ensues, the quality of the collaboration is one of the matters which will come under examination. The client is commonly assumed to have a right to expect collaboration among all parties concerned of the quality necessary to achieve a satisfactory end result.

Some aspects of the division of liability should be clear enough. Liability for the reinforcement design, the quality of the concrete, the factory inspection of component manufacture, the structural design of the support and fixing system and for the handling and site assembly lies naturally with the structural engineer, while the architect is normally responsible for deciding on the insulation, the ventilation of the cavity, the jointing technique, any drainage or damp-proofing required, fire risk control and, of course, the appearance of the products on the finished building. But there must also be a good interchange of advice and questioning all along the line if success is to result.

4.4.12 Alkali/silica reaction

Most designers know by now about damage to concrete by alkali/silica reaction (ASR). Briefly, the mechanism of deterioration is that an interaction takes place between alkaline pore fluids originating in Portland cement and reactive minerals in a few types of aggregate. The result is a gel which imbibes water and causes a disruptive volume expansion in the concrete, sometimes recognizable by random ('map') cracking. The cracks may exude some jelly-like material. Not all cracking of this kind signifies ASR and a laboratory examination of a sample is needed to establish it.

The reaction is only likely where there is a combination of high cement content, high alkali in the cement, a susceptible combination of aggregates, and wetness. The last points to areas of high rainfall or situations where there is persistently severe condensation as being particularly at risk, but bridges and highway overpasses seem to have been exceptionally vulnerable, perhaps aided by persistent wetness under the road surfacing material.

The BRE has publications giving guidance for avoidance and up-to-date information on the progress of research. It is an active field of study.

4.4.13   Summary

Among cavity wall-cladding systems reinforced concrete ranks close to brickwork in weight and thermal mass, but unlike brickwork it has very little fortuitous ventilation and deliberate provision is necessary for the cavity, or a classic thermal pumping situation can be created.

The manufacture and fitting of pre-cast concrete cladding is a skilled operation – a combination of craftsmanship and science-based technology – and it should only be entrusted to firms able to demonstrate that they have these attributes.

The design of the cladding, the supervision of its manufacture, its safe transport and its fitting on site are matters which require good rapport and collaboration between the architect and the consulting structural engineer.

Concrete quality depends greatly on the types of aggregate and their size range in the mix and on the water/cement ratio of the bonding paste. These and related factors determine the resistance to carbonation and therefore the depth of cover needed for reinforcement for the durability of components, because when carbonation reaches the steel, its protection against corrosion will rapidly disappear.

Some key points to note are as follows:

- Carbonation takes place more rapidly from the inner face than the outer, and the cover for steel must be at least equal on the two faces as well as along edges of components.
- Concrete tends to develop cracks at sharp inner corners due to shrinkage stress concentration.
- Tiling and mosaic on concrete pre-castings prevents drying from the treated face, so drying shrinkage concentrates on the back and causes bowing.
- Natural or artificial stone finishes have to be clipped in place securely.
- Artificial stone is so greatly a matter of skill for successful manufacture that one should see examples of a firm's work which has been exposed for some years.
- The trueness of size and shapes of pre-castings is critical to reliable assembly. Tolerances have to be small.
- Mastic seals must be used for vertical joints to avoid the collective expansion risks if mortar is used.
- The manufacture of components and their safe delivery should be rigorously supervised. Do not accept damaged components on site, even if repaired.
- The respective design and supervision liabilities of the architect and the structural engineer should be sorted out in advance.

4.5   Rendered finishes for cavity systems

Rendered finishes over cavities containing insulation have become popular as a way of improving inadequately insulated domestic property but it is also a valid technique for new construction. Its weight and thermal mass as an outer cladding put it between thin stone and the sheet finishes to be discussed in Section 4.7.

The rendering is applied to stainless steel metal lath, and in accordance with cavity requirements described in Section 4.2 there should be an air gap between the rendering and the insulation for ventilation. Some remedial work in Britain has omitted the air gap and the risks are discussed below.

4.5.1   Metal lath as the base

The metal lath is framed out the necessary distance from the inner leaf. Stainless steel lathing is a must and it should be continuous over its support ribs so that there are no breaks in its continuity as reinforcement. Discontinuities lead to shrinkage cracking in the rendering. Sheet-to-sheet continuity must be ensured by overlaps secured together.

4.5.2   The rendering

Traditionally, renderings on lath comprise three coats – the familiar 'render, float and set' trio. Pre-mixed, bagged material is now commonly used and is likely to comprise cement, lime and sand in such proportions as 1:3:9. Ideally the successive coats should each be a little weaker in a progression such as 1:3:9, 1:3:10 and so on, so that when the relatively stronger preceding coat has set, its strength can better restrain shrinkage in its follower. However, proprietary pre-mixes are not necessarily available in such a succession. At least the successive strengths should never get greater and preferably the outer coat should be appreciably weaker. A straight cement/sand mix should never be used; it is too harsh and shrinkable.

Each coat must provide a strong mechanical key for the next, mainly for the secure adhesion of the whole multi-coat assembly but also to exercise shrinkage restraint on successive coats. It has been usual to specify that each coat of fresh mortar should be 'scratched' to provide a key for the next but this often results in nothing more than useless grooves. It is absolutely essential to undercut the surface generously and raggedly if a scratched key is to provide any worthwhile integrity for the undercoat system and it should be watched carefully in inspections.

The sand should be well graded as described for mortar in Section 4.3, and undercoats should never be applied in thicknesses greater than about 10 mm; otherwise they can get so heavy that they can pull away before setting properly and are then unable to bond well. Feather-edging should also be avoided – it sometimes gets done over uneven backings – because thin render can dry out too quickly and never have any strength; it is especially at risk in hot weather and notably so in sunshine. If thinness is necessary anywhere, the area concerned should be protected and kept damp.

If mixing is done on-site and a plasticizer is used in place of lime for workability, it is absolutely essential to avoid an overdose because it bubbles and prevents adhesion, as remarked about mortar.

In Britain it is usual to apply undercoats and the final finish by trowelling. In continental Europe, where renderings are much more reliable than in the UK, a different technique of application is used. The operative usually holds the material in a large shallow scoop, picks it up

Figure 4.28
A Continental rendering on a British building in Welwyn Garden City.

on a big heart-shaped trowel and throws it on with an easy, scooping motion. The force is sufficient to dispel the films of air that trowelling often locks in and which reduce adhesion, and the thrown material automatically provides a rough surface to which the next thrown coat adheres well without needing to be scratched. It is successful but has rarely been accepted by UK operatives despite great efforts in the past by BRE to get it adopted.

Figure 4.28 shows a building in Britain with a continental European rendering, Roche Products in Welwyn Garden City by Otto Salvisberg of Zurich, built in 1938. Its rendering, is in excellent condition nearly 60 years later. The pre-mixed materials were imported.

For the finishing coat, unless it is to be a stucco or roughcast, the aim should be to have it uniformly absorbent and sufficiently weak to ensure that the bond to its backing coat can prevent shrinkage cracking. Roughcast usually avoids cracking by distributing shrinkage microscopically around the coarse aggregate when it is thrown on.

Assuming that the adhesion of a rendered finish is satisfactory, that its aggregate is well graded and that the undercoats have stopped moving significantly, any remaining risk of cracking arises mainly from the trowelling habits of UK operatives for the final coat. They often still aim to achieve a smooth finish and try to get this by trowelling with a wood or even a steel float, but this works a cement-rich laitance up to the surface which reacts at once with atmospheric carbon dioxide to form a skin, and being exceptionally rich in cement, it is both strong and shrinkable and commonly results in crazing both the skin and the still-underset material beneath. This happens in the first 15 or 20 minutes, especially if there is hot sunshine on the render, but it is usually hard to see at this stage. After a few hours it becomes more visible and some water thrown onto the surface will show it up well by the rapid absorption that takes place along the craze lines. The only sensible course then is to have the finish coat stripped off at once and redone. It will never look well, and the cracks that eventually develop along the craze lines can sometimes cause trouble. If paint is applied to mask the crazing, the latter will win and the paint will fail along the craze lines.

The Continental practice for finishes avoids these risks. The top coat, like the undercoats, is thrown on but then, 15 or 20 minutes later when the initial set has taken place it is vigorously scraped to a flat finish, removing as much as 40–50 per cent of the coat. No trowelling is needed, no laitance or crazing is left and good surface absorption and a uniform texture results that can weather smoothly. Regardless of whether the finish is thrown on or trowelled, it can and should be scraped.

The scraping is often done with a long, straight steel rule but European operatives also invent their own hand-sized scrapers, sometimes simply saw blades set diagonally in a wood trowel, and they often also put slightly lumpy aggregates in the top coat so that the scraping will tear the surface to form coarse textures. It is unfortunate that pre-1939 efforts of BRE to get Continental application practices adopted in the UK were not successful; handsome, crack-free renderings are the rule in France and Germany. It might be rewarding to mount a fresh technology transfer exercise.

4.5.3 Implications for design

The fact that these rendered finishes are mainly used on domestic and other low-rise property where the passive loss of indoor water vapour through the inner leaf should not be discouraged implies that if the void enclosed by the rendering is not ventilated, persistent condensation should be expected on the inner face of the rendering which will keep the void humid and the insulation damp. By this logic the air gap should be vented top and bottom, although the vents need not be large. It has to be said, however, that there are no known reports of such cavities being opened up to see how much dampness actually accumulates. Some will escape through the rendering, though slowly. It needs investigation, especially to establish if insulation is retaining its value.

A second implication is that the load-bearing wall for the low-rise construction then becomes simply the inner leaf and logically can revert to being of solid rather than cavity construction in new buildings. This would restore firmness to what has often now become a rather flimsy construction, with a high thermal mass providing thermal carryover which can be exploited to advantage by high-inertia central heating for the benefit of indoor comfort.

4.6 Stone veneers

The stone industry was concerned to discover a new market after 1945 when thick carved stone largely disappeared from the architectural scene. Its most cost-competitive form was in thin sawn sheets, and happily for the industry this consorted well with the architectural ideas that were developing. Flat veneers were in. As a cladding system its weight is typically from a quarter to a third that of brickwork or concrete and so is its thermal mass.

Stone cladding is typically done with rather small slabs in two or three lines per storey height, although larger panels, sometimes storey height, are occasionally used.

4.6.1 Stone types

Not all types of stone have proved suitable; granite, limestone, marble, travertine, sandstone and slate have gradually established themselves as the most satisfactory for the purpose, but all can misbehave sometimes. One is habituated to think of stone as a stable and well-behaved material but it is now appreciated that cutting it thin alters the physical behaviour and character of some types and can limit their usefulness. It makes all types more responsive to temperature change by reducing their thermal inertia, and the responsiveness is quickened and enlarged if there is insulation in the cavity. Marble, sandstone and slate have the largest thermal movement coefficients and in the thinnest cuts, 25–35 mm, they will experience seasonal temperature changes of as much as 60° or 65°C from about −10°C or less, up to at least 50°C. Colour is, of course, a factor in heat absorption and expansion. Allow therefore something like 0.6 mm/m of movement for these stone types; granite will move perhaps two-thirds to three-quarters of this amount, and limestone about a quarter.[5]

Granite

A polished finish to granite provides useful protection by sealing pores. The 'polish' is not a coating but a very fine grinding.

Both bush hammering and thermal finishing by flaming reduce the thickness by about 3 mm. Flaming also causes microfracturing of embedded quartz and feldspars and this allows water to be absorbed to a depth of 5 or 6 mm. Feldspars are subject to cleavage and may be surface-fractured by bush hammering due to cleavage in the particle structure. These finishes therefore reduce the bending resistance and elastic deflection under wind forces, and there is a gradual further degradation by freeze–thaw cycling. A 30–35 mm thickness of granite may lose 20 per cent or more of its nominal bending resistance and perhaps 30–40 per cent of elastic deflection. The thicker the stone, the smaller the proportionate loss will be.

A recommended safety factor for structural computations is 3.[5] For large panels exposed to strong winds a thickness of 50 mm is suggested where a polished finish is used, and up to 75 mm or more where bush hammering or flaming is used.

Marble

Marble is a particularly variable stone, sometimes excellent but sometimes hopeless for exterior cladding, with the highly decorative Breccia types especially doubtful. Marble has a calcite component, and in warm weather particles near the face may interlock so that a fraction of the thermal expansion is retained after each heating cycle. By this process, known as hysteresis, the exposed face of the stone gradually acquires a permanently larger volume while the cooler material behind is less affected, and the panel slowly takes on an outward bow. The thinner the stone, the greater is the effect, because thicker material can offer more restraint by reason of greater panel stiffness and a larger proportion of cooler stone behind the face.

The bending effect can be serious and a well-known example is the 82-storey Amoco building in Chicago, which was clad with some 47 000 panels of white Carrara marble in panels about 1.2 m square and

30–35 mm thick. These gradually bowed outward by as much as 35–38 mm and eventually a decision was taken to replace all the marble with white granite. Considerable research was done during the investigation by the consulting engineers.[5]

This thermal dilation of the face of the marble increases its porosity and makes it more vulnerable to attack by sulphurous and sulphuric acids and freeze–thaw cycling, and it then begins to granulate quite readily under light impact forces such as pelting by wind and rain. The fine-grained, relatively pure marbles are apparently the most likely to suffer degradation in these ways. Experience suggests that granulation may also be a feature of stress fracture and cyclic loading generally.

An evaluation of any thin cut marble should therefore be made before use. Tests should include cyclic heating and wetting to check for bowing risks, and because the modulus of rupture of stone can be affected by freeze–thaw cycling, this should be checked after the other tests have been made on the samples.

Some marble gets formed under immense pressures in the earth and a gradual volume increase may take place when it decompresses after being cut free in the quarry. The amount cannot be large, but it underlines the importance of adequate allowance for thermal and other movements.

Marble strength diminishes after an initial period of heating–cooling and freeze–thaw cycling and accordingly a safety factor of 5 has been recommended for strength computations.[5]

Limestone

Like its cousins the marbles, limestone is vulnerable to slow degradation by atmospheric acids and to a lesser extent by carbonic acid and ammonium salts. The sulphur-based acids form gypsum and the others dissolve the lime component, causing a loss of material. A chemical delaying treatment is possible which decreases the permeability of the surfaces and is not affected by these aggressive chemicals. Coatings sometimes applied to discourage graffiti should be used with caution however: they can lock in moisture which thereby encourages frost damage and other exfoliation effects. Any micro- or macrofracturing in the stone gives deeper access to rain and water vapour and if it carries in any of these chemical pollutants, degradation will quicken a little.

Travertine, slate and sandstone

It seems that less is known about travertine, slate and sandstone in thin cuts than about the stones mentioned previously, but a few comments can be made.

Sandstone and slate are among stones that come from rock deposits which have jointing systems, i.e. planes of imperfect bond that occurred during formation. Travertine was formed as a calcareous deposit. The joints are familiar as bedding or cleavage planes or banded concentrations of minerals, and they are planes of weakness that may come unstuck under structural or thermal stress. With slate this characteristic is exploited to make roofing slates, but no one wants wall cladding to turn into roofing material *in situ*. Cladding slate has to be carefully chosen.

As with slate, but on a larger scale, the cleavage planes of sandstone have been exploited traditionally to provide large slabs for stone roofs, typically set in 35–50 mm thicknesses, but again these weak planes must not occur in wall cladding or they may shell off *in situ*. Careful selection is necessary. Cleavage can be particularly dangerous at fixing points.

Travertine does not have planes prone to separation, but occasionally elongated voids occur that will weaken a slab undesirably for cladding and a watch should be maintained for signs of them at the quarry and on-site.

Chemically, sandstones are either oolitic or siliceous, i.e. bonded respectively by calcium or silica. The latter resists degradation well, but in the oolitic sandstones the binder is calcareous and therefore can be attacked by the same agents as those to which limestone is vulnerable. Very substantial erosion can develop – Birmingham and Chester Cathedrals are examples – and can be compared with the durability of the siliceous sandstones used widely in the north of England and in Scotland.

Some sandstones are surprisingly porous and may become visibly damp for quite long periods. This may become troublesome in freezing weather.

4.6.2   Artificial stone

Artificial or cast stone is a specialized form of concrete used as veneer cladding. There is a British Standard for its manufacture (BS 1217: 1986) and it is important to ensure that any such material used for cladding conforms and has the manufacturer's written assurance that it is suitable for the proposed use and method of fixing. It seems generally to be more permeable than its natural counterpart, which is not too surprising, and unless the manufacture and curing are of at least BS quality, crazing and other faults may develop.

The artificial product is not necessarily less durable than the real thing and can sometimes be more so. Fitzmaurice[6] describes a find of artificial stone by Viollet-le-Duc in a survey of Carcassonne, made and put in place in repairs to the Visigoth walls in the twelfth century. The walls were mainly of sandstone, but the artificial product had proved more durable than the natural. BRE obtained some for analysis and found that it comprised large fragments of sandstone cemented by lime and crushed potsherds, i.e. a pozzolanic mortar. It was in excellent condition.

4.6.3   Fixings

Fixing practice as it developed after 1945 typically comprised projecting brackets engaging grooves cut in the bottom edges of the stones to take their weight, or uptilted brackets engaging sloping grooves cut into the backs of the stones near their bottom edges. Side clips engaging grooves in the edges of stones near the top held them back against spacer pads of mortar to transfer to the main structure the push and pull of pulsating wind forces.

When load support brackets and side clips engage adjacent stones at joints, matching grooves facilitate accurate lining up. The brackets need three-way adjustability, but the early types were rather primitive and not always very reliable. A recommended safety factor for anchorage design is 4.[5]

The mortar pads have largely disappeared with the advent of insulation in the cavity, and this, together with the ventilation gap, has meant

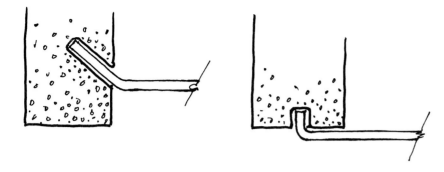

Figure 4.29

that brackets and clips have a bigger distance to span so that it is now necessary to make a comprehensive structural check on the whole panel and support system. The following are points requiring attention:

- The ability of the thin-cut stone to retain its integrity when receiving and transmitting wind forces to the clips and brackets.
- The stone will have been thinned by cutting the fixing grooves. The cuts should be made with all edges and corners rounded to avoid stress concentration, and the remaining stone around the grooves must be able to stand up to wind pulsing on the fixings[5] (Figure 4.29).
- The clips and brackets must be designed and fitted so that wind forces are transmitted as nearly as possible simultaneously. Otherwise, individual fixings may have to take the full loads for a moment and be overstressed.
- The brackets must carry the cantilevered panel loads with an adequate factor of safety. It is doubtful if any line of stones should any longer be expected to carry the next line or lines above them.
- The adjustable elements of three-way fixings must all accommodate the wind and dead load stresses.
- The system must contain no element subject to the risk of premature failure by fatigue over the expected life of the cladding.

Stones containing feldspar crystals may be weakened at grooves or tie positions because of their cleavage characteristic. The stresses involved in dry sawing or drilling can cause microfractures that can propagate from the cleavages. They may travel some distance, and if the crystals are large and the stone sufficiently thin, a critical loss of strength may occur at the anchorage.

Clips and brackets must never engage the holding grooves by jamming or grouting. This can cause bending moments to be imposed on the stone due to deflection or shrinkage of the main building structure and these can cause cracks that will propagate, sometimes eventually crossing the panel from one fixing to another. This is very dangerous.

Despite such structural checks it is evident that a good deal of the safety in thin stone cladding depends on the skill and care of operatives. It is far from fool-proof and risks are such that improved proprietary methods of fixing are being marketed. One line of development has used subsidiary frames to hold the individual stones, with the frames fixed to the main structure. The designer should discuss the available options with stone cladding specialists before design commitments harden.

4.6.4 Joints

Joints are normally mastic-pointed. They should be checked before pointing to ensure that there are no spacers or hard objects left in them which could become pressure points and cause cracks or spalls when the stones expand or get shifted by structural movement.

Since it can be assumed that leakage will develop at joints because adhesion of mastics is never perfect and gradually degrades, the clips and brackets must not be able to let water cross from the cladding to the inner leaf, nor should they be subject to corrosion. All mastic must be accessible for replacement.

The following case exemplifies the problems that have to be overcome in the cavity behind the cladding.

Case note

An office block was clad in thin black marble. Windows were metal framed and flush with the cladding face. Lintol linings inside were decorative panels and water began to leak through the stone joints and into window frames, emerging at a variety of embarrassing locations. Openings in lintols and reveals showed that strong upward air currents pervaded the uneven voids between the inner structure and the cladding, evidence that there were no fire breaks in the cavity. Short of dismantling the whole wall, which was not an available option on this occasion, all that could be done was a high-grade exercise in resealing the mastic joints and the placing of drainage outlets at the base of the cavity.

It is not practicable to produce satisfactory generic details to provide fire breaks, drainage, ventilation and damp-proofing over door and window openings without introducing considerable complexities into an otherwise essentially simple architectural concept. There would need to be much discussion with stone cladding specialists, probably on an *ad hoc* basis related to a specific building project. The alternative is to look to the open-joint technology described in relation to rainscreen cladding.[7]

4.6.5 The learning curve about thin stone cladding

Four cases illustrate how the learning process about thin stone cladding has developed since 1945.

Case 1

Initially no allowance was made for the shrinkage and deflection of reinforced concrete framed buildings. On one such building clad in marble the frame was badly misaligned and brackets had to be shimmed out as much as 70 or 80 mm in places. Even then they could not always engage properly the grooves intended to receive them. Thermal cycling, concrete frame shrinkage and traffic vibration eventually caused disengagement and a slab fell into the street. Other stones were then found to be insecure and a general refix had to be undertaken.

Case 2

In this case the stone was polished granite. The brackets engaged the slots but often quite tightly, even sometimes having grout in the slots, and shrinkage and deflection of the concrete frame gradually applied bending moments via the brackets. These initiated fractures which gradually propagated, aided by thermal cycling. The fractures were often difficult to see and had to be traced by dyes. Some were found to have travelled across the full width of stones and many were well on their way. The jointing between stones was entirely by mastic, and this held the fractured stone in place. None fell, but all the cladding had to be replaced. Stainless steel was chosen for the purpose.

Case 3

Here Portland stone in 75 mm thickness was used in continuous vertical bands on a 10-storey building. The supporting brackets were horizontal, of mild steel, with a 90° turn-up at the outer end to engage grooves in the bottom edges of the stones, carrying every third stone. The concrete frame shrank, probably about 25 mm overall, but not the stone, of course, and the brackets on the upper storeys were drawn down, pressing on all the stone below. Some brackets disengaged, some were bent, and a few were broken. Much of the stone had lost direct connection to the building frame and was loose. A dangerous situation existed but fortunately no collapse occurred (Figure 4.30). In a difficult remedial programme all stones were fixed back by bolts.

Case 4

Big panels of slate up to about 1.6 m long delaminated seriously and one panel broke off, dropping into a stair much used by the public. Fortunately, no-one was injured.

Figure 4.30
One of the fractured clips.

Liability for falls

Stones that fall are a huge danger to the public and the designer may be held liable if an accident occurs. If the stones develop curvature and/or the surface deteriorates and becomes ugly, replacement may become necessary at great expense and again the question of liability may arise. It is a cladding system to be treated with the greatest respect and care, especially if it is used at any significant height and most particularly when fixed to a reinforced concrete frame. Do thorough analyses and adequate testing before selecting any proposed stone.

4.7   The rainscreen concept

In Section 4.2 it was explained that water-tightness is not practicably attainable in outer-leaf jointing and that it is not a realistic or worthwhile goal. Leakage will take place from the start or will eventually develop by degradation of jointing material or hygro-thermal loosening of the jointed units. Ventilation is needed as a drying agent and also to ensure that water vapour pressure from indoors does not build up in the cavity but escapes to atmosphere. The outer leaf, as remarked in Section 4.2, is best seen now simply as a weather screen protecting insulation in the cavity.

4.7.1   Development of the concept

The fullest recognition of this idea is found in modern so-called rainscreen cladding in which the joints are open and rely upon quick wind-pressure equalization in the cavity to prevent the entry of wind-borne rain. A brief outline of the history of its development will help to explain the way the concept works. A more complete description has been provided in a report by Anderson and Gill[7] and by other literature referenced there.

As with most ideas in building, this one has early origins. Tile hanging is an example and another is open-jointed vertical boarding which was used on Scandinavian houses. However, the first modern research reference drew only obliquely on such traditions when in 1946 Johansson, studying the increase of heat conductivity due to water absorption in brick walls, said that it would be a great step forward if an outer, water-repelling screen could be fitted to keep walls dry. It was still quite a long way from the concept of an outer leaf as a screen for a cavity in which insulation is given sheltered accommodation, but it was a forward glance.

The next step appears to have been taken in 1952 by the American architects Harrison and Abramovitz in their 30-storey Alcoa building in Pittsburg, where they used storey-height aluminium panels with baffled rather than sealed joints in association with back drainage. When examined 20 years later it was found still to be in good condition.

Drained and ventilated screens of this kind were then studied by BRE and at Princeton University, but the idea of pressure equalization awaited a publication by Birkeland[8] of the Norwegian Building Research Institute in 1962, in which he identified wind-induced air pressure differentials as a main cause of rain being pushed or carried through all the usual types of joints. He argued that a good way of preventing it was to design a screen with joints so open that negative and positive pressure surges from gusting wind were equalized almost instantaneously inside the cavity and out, so that inward carriage of rain by wind would not often take place, or would be minimal.

The next development followed quickly, in 1962, when G. K. Garden in Canada[9] showed that in order to get instant pressure equalization in the cavity it had to be divided into compartments, because gusting is localized on facades and the pressures created in the cavity had to be prevented from flowing too far from the pressure entry points or the necessary compression would not take place. He found that the size of the compartments should vary across the face of the building, being quite large in central facade areas but relatively small near the corners where the rate of wind pressure change is greatest, and with corners blocked absolutely to prevent air escaping around them to feed suction positions. Subsequently he and a colleague, Dalgliesh, put these recommendations into sharper focus, saying specifically that:

- Vertical closers should be provided at outer corners and at about 4-feet intervals for about 20 feet from the corners.
- Horizontal closers should be placed near the tops of walls.
- Vertical and horizontal closers should occur up to 30 feet apart in other wall areas.
- The closers do not need to provide a complete seal but should give good resistance.

These have been redefined by experience and research, and now stand as follows:

1. Closers are needed not only at the outer corners of the main shape of a building but at outer corners also of large recesses or projections and at the top of the building.
2. Vertical closers should be spaced 1.2–1.5 m apart for a distance from corners of about 25 per cent of the lesser plan dimension.
3. The remaining vertical closers should be no more than about 5 m apart and the horizontals about 10 m unless the cavity contains materials fire-graded below Class O, in which case a reduction to 7.5 or 8.0 m should be made.

Gusting wind imparts a horizontal kinetic force to raindrops, so that even if entry of the wind itself is arrested at an opening by the compression of the cavity air, the drops have enough momentum to carry on inwards a little. The opening therefore needs to be a limited labyrinth, shaped so that there is easy passage for air while the drops are stopped and drained outward. Anderson and Gill describe a number of possible shapes for the horizontal and vertical joints (Figure 4.31).

These are the bare bones of what has become an established although still-developing technology, fleshed out by experience and research and now embodied mainly in proprietary systems and the expertise of specialist consultants.

4.7.2 Points of detail for design

The fact that the idea got initial usage from a perceived need for a way of improving leaky and poorly insulated walls gave rise to the concept

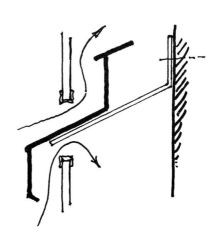

Figure 4.31

of low-weight screens of such materials as sheet metal or GRP, and this inhibited for a time the thinking necessary for the extension of the open-joint technology to the heavier and more durable-looking finishes needed for a place in the general vocabulary of architecture. Thin stone panelling was the first of these heavier materials to be used with open jointing, and concrete cladding seems a possibility.

A few points of technique:

- Horizontal open joints must not be too narrow. Below 5 or 6 mm they can easily be blocked by rain running down unabsorbent finishes; 10 mm is a recommended minimum and some are as wide as 15–25 mm in practice.
- In an oceanic climate given at times to severe combinations of wind and rain it is sensible to protect the horizontal open joints with a drip lip.
- In order for pressure equalization to work, the compression of the cavity air must not be allowed to dissipate. Research has shown that the inner wall receives pressures about the same as those that the gusting wind applies externally, and it must therefore be made reasonably air-tight. The compartment closers and the door and window reveals are less critical, no doubt because some of the external pressure in a gust of wind gets applied to each side of them, but they must offer reasonable resistance. Densely packed rockwool is one of the compartment closers in use.
- Fixings for the screen are generally to vertical rails of some kind, usually designed on their outer edge to be vertical open jointing, and sometimes they are also designed to be cleaning cradle guides.

At the time of writing the learning curve about lightweight screen cladding still continues upward, and a failure of cladding fixings which occurred on a remedially applied rainscreen is instructive.

Case note

The application was to high-rise flats, the original walls of which were defective. The overcladding was designed for the job and part of it was open-jointed, designed for pressure equalization. Some of the panels blew off fairly soon.

The fixing system was complex. Threaded anchors were set in the original structure to provide primary fixing and steel plates were fastened to them by nuts. Stainless steel strap brackets were fixed to the plates by nuts and bolts and some had to be shimmed out so far that nuts did not fully engage the bolt threading. Vertical rails were bolted to the strap brackets and the panels were then bolted through spacer blocks to a thin aluminium angle, which in turn was bolted to a thicker metal strap through a rubber-type gasket to an aluminium extension on the vertical rails. It was all needlessly complicated.

Rubber washers were incorporated at all the bolting positions so that in the end the fixings were found to comprise seven variably flexible components with eight layers of rubber. The nuts were not locked, and when the panels came off it was found that some nuts were undoing themselves and others had already disappeared. The rubber washers had deformed or split or were missing, and bolt holes showed severe deformation. On analysis the rubber proved to be of a type sensitive to degradation by atmospheric ozone.

Computations indicated that the flexibility of the panels and fixings allowed a screen deflection of the order of 10 mm under gusting and wind loads. In principle, this would lead to much compression and decompression of the washers, not only damaging them but at the same time causing unlocked nuts gradually to unscrew themselves.

The lessons are clear. Keep fixings simple and use types that will not undo themselves. Avoid rubber-type flexibility and provide for any necessary tolerances in some other way.

4.7.3   Tile hanging

Traditional tile hanging has some resemblance to rain-screening. It is certainly open-jointed and it excludes rain by water shedding as well as by some inadvertent pressure equalization. It works excellently, exactly as tiles and shingles do on pitched roofs. If it has not already been used with insulation in the cavity behind it, no doubt it will be. It would appear to present no difficulty, and one can envisage, for example, attractive and effective traditional combinations of upper-storey insulated tile hanging with insulated cavity brickwork for the ground floor.

4.7.4   Sound insulation

As explained earlier, open jointing necessarily deprives any kind of outer leaf, light or heavy, of most of its sound reduction potential, and the selection of the inner leaf will have to provide all but perhaps 5 to 8 dB of the noise protection of the wall.

4.7.5   Summary

Although open jointing and pressure equalization have been closely related to thin sheet cladding in the development of the rainscreen concept, they are not interdependent. The initial attraction of lightweight cladding with a simple jointing system that avoided the sophistication required for waterproof joints was as a means of improving conventional walls without adding greatly to structural loads, but the open joint and pressure equalization are equally valid with heavier screens.

4.8   The inner leaf

4.8.1   Design decisions

The parameters of inner-leaf selection are mainly structural, thermal and moistural. Structurally, is it to be load-bearing or wind-resisting? Does it have to help stabilize the outer leaf? Thermally, is it to be of high thermal mass and low insulation or to have some other balance? Moisturally, is it to be air- and vapour-tight or permeable to one or the other, or both? It should be reasonably stable dimensionally and it may have to provide good sound insulation. Design decisions about cavity wall systems must deal with these inner-leaf questions.

Where brick and block cavity construction is used for low-rise work, the inner leaf usually takes the vertical loads but gains some of the stability it needs by ties connecting it to the outer leaf. The horizontal wind loads are received by the outer leaf, which gains stability by its ties to the inner leaf.

When the scale of construction moves up to that of multi-storey framed construction the frame takes the vertical loads, but because the outer leaf will usually be divided into areas of limited size with soft joints to accommodate thermal or moisture movement, they will depend upon connection to the inner leaf for a lot of their stability under wind loading. The inner leaf must therefore be tied to the structural frame to be sure of having the necessary stability itself. It will also usually have to be of brick or dense concrete blockwork for strength. Structural engineering advice should be obtained.

Where the outer leaf is of storey-height pre-cast concrete units or of cladding carried on mullion framing, the wind loads on the outer leaf will be taken direct to the primary frame, but because wind pressure builds up in the cavity by reason of air leakage or ventilation openings, the inner leaf must still be able to take a considerable share of the positive and negative wind loadings. Where pressure-equalized rainscreening is used, the inner leaf will receive full wind loading, as already explained.

Where thin stone cladding has to be carried by the inner leaf, the connection system is different and the inner leaf has to be able to take both its vertical and horizontal loadings.

4.8.2   Ties and mortar

Since wall ties have to act both in tension and compression, mortar strength must be carefully watched for both the inner and the outer leaf. There have been some dangerous failures.

4.8.3   High-rise load-bearing brickwork

Brickwork walls 225 mm thick and built of high-strength bricks and mortar are sometimes designed to be load-bearing for heights of as much as eighteen or twenty storeys. Usually they are used for flats if they are used at all, and they rely for much of their stability on the carriage of concrete floors, thereby forming a cellular structure usually able to resist all likely wind loads. Such load-bearing walls would form the inner leaf of a cavity system, but oddly and rather irrationally they have been used sometimes as the outer leaf. In high blocks the exposure is severe and being of very dense bricks and mortar they usually leak badly. Also, their DPCs are under exceedingly high pressures and if they contain plasticizers they can suffer loss by migration. The floors, of course, would act as cold bridges. The proper place for high-rise load-bearing brickwork is always as the inner leaf.

4.8.4   Timber-framed inner leaves

The combination of a timber-framed inner leaf and a brick veneer is not widely used in these islands, but happens occasionally for low-rise property and problems can occur. The difficulty is that while the brick veneer is expanding to its equilibrium moisture content, which takes several years, the timber frame will be shrinking to its own long-term moisture level, which it will do in one or two seasons and then remain to some degree seasonally cyclic around that moisture content level. Initially the vertical differential can be expected to be as much as 8–12 mm at first-floor cill levels, with another 2 mm or so at eaves level in a two-storey dwelling, but the figures depend on aspects of timber frame design discussed below. The differentials develop laterally as well as vertically but are likely to be of a lower order.

The reason lies largely in the way timber house framing is done and to some extent on the timber used. Five factors are significant: the initial moisture content of the timber; the type of timber used; the eventual in-service dryness; the amount of cross-grain timber in the support system; and the uniformity of support at the base for the loads involved.

Initial and in-service moisture content

Recommendations have been made by various authorities to restrict the moisture content of delivered timber to 3 per cent above the expected long-term equilibrium figure and to give it sheltered storage on site to hold it at that level or less until put in place. It is certainly desirable to start with reasonably dry timber and to try to keep it so, but its moisture content as built will nevertheless be determined by humidity values in the open air during storage and the weather during construction, as well as by the dampness of new-built brickwork. Measurements have shown that the as-built moisture content can be expected to be as high as 20–22 per cent while the later equilibrium figure will depend on the indoor climate maintained by the occupiers. If RH values are reasonable and heating is intermittent, it will drop perhaps to around 15 per cent, while with continuous winter heating, 9–12 per cent (or even lower with floor heating) is likely to occur for a time, rising again in summer when indoor RH values become similar to those outdoors.

The wood

Framing is done with common commercial softwoods such as Scots Pine, Western Hemlock, Douglas Fir or Spruce. One can get data about the moisture movement values of these and other woods if necessary, but of these four, Scots Pine has the highest figure, Hemlock the next, and the other two the lowest – about three-quarters of the higher figures. The range is not large.

Timber does not shrink or lengthen very much along its grain with moisture content changes but does so across it, the latter depending on how it is cut from the log; tangential moisture movement is roughly twice as great as radial. A high radial grain content is obtainable only by quarter-sawing the logs, but this is expensive because of high wastage and handling costs. Board sawing is cheaper and customary for framing timber, with a corresponding variation in radial grain content according to how far the cut board is from the log centre. Commercially grown timber has greater movement than natural growth, but the latter is becoming less and less available and more expensive as dwindling world resources are cut and used.

The framing

The practical implication of this is that the amount of long-term shrinkage in a frame assembly will depend on how many cross-grain timber elements are designed into it across the direction of movement, vertical or horizontal. Vertically the framing studs will not shrink significantly; it will be the horizontal sole plates, wall plates and perhaps the joints that do most of it. The sole and wall plates are unavoidable, but sometimes they are doubled and this adds pointlessly to shrinkage.

Whether joists play a part in the overall vertical movement depends on the type of framing used. Platform framing brings them fully into the wall system, while balloon framing reduces their relevance (Figures 4.32 and 4.33). In countries where experience of brick veneered timber framing is greatest, balloon framing is usually recommended.

Support at the base

Lastly there is the influence of non-uniform support at the base. If it is uneven, as concrete often is but shouldn't be, the loads will concentrate at the high points, adding compression to shrinkage at relevant positions in the cross-grain elements. Deformation and creep will then cause a slight overall downward movement until stresses are redistributed.

These differential movements can be very troublesome collectively unless they can be kept to somewhat lower figures than those mentioned. Roof loads may be transferred from the frame to the outer cladding, and door and window openings in the frame can get unpleasantly displaced from those in the outer leaf. Substantial cracks can open up, distortion can occur, and damage can be done to the cladding where it and the frame are connected, and these troubles may recur seasonally.

4.8.5 Thermal factors

Where a cavity contains good insulation, a high thermal mass inner leaf can act as a thermal flywheel, needing only relatively small and infrequent inputs of heat to keep its temperature fairly constant but taking longer to warm up from cold, or the reverse, whereas a low thermal capacity leaf will be quicker to warm up or cool off and therefore needs more continuous heat inputs and perhaps a bigger plant capacity to give similar thermal stability indoors. A good thermal analysis should be done to decide which is the better buy in any given case. In museums, art galleries and other buildings where thermal and humidity stability is exceptionally important for conservation, a well-insulated high thermal mass is valuable in itself and is an insurance against undesirably abrupt shifts of the indoor climate when the weather changes or a loss of power occurs. (See Chapter 2 for the main discussion of museum and gallery envelopes.)

4.8.6 Air- and vapour-tightness

In Chapter 2 a distinction was drawn between air-conditioned buildings where the envelope should be air- and vapour-tight so that indoor climates can be closely controlled, and buildings not air conditioned where indoor/outdoor vapour pressure equalization should be allowed to take place passively through the envelope. To this design point

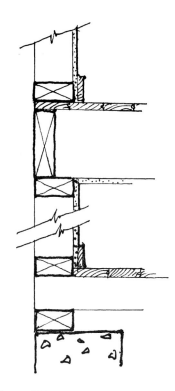

Figure 4.32
Platform framing.

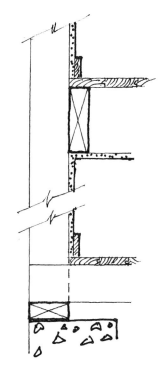

Figure 4.33
Balloon framing

another must be recalled to which reference was made earlier (page 122), namely that if open-jointing is to be used in the outer leaf for pressure equalization in the cavity, the inner leaf has to be reasonably air-tight to prevent pressure loss.

When the inner leaf has to be vapour-tight it will usually need a good vapour control membrane, but if this has to be applied to the cavity face of the inner leaf any significant air leaks must first be sealed. Air conditioning is always designed to pressurize the interior atmosphere and local air leaks can push off the vapour check or open up a lap. The surrounds for door and window openings are particularly leaky positions.

In North America, timber framing for house construction is usually stabilized by plywood sheathing. However, there is a point to be watched about using it in our oceanic climate because plywood is both air- and vapour-tight and will prevent vapour loss to the cavity. It should be perforated in each bay of the frame for this use in oceanic climates.

4.8.7 Pattern staining

Pattern staining of indoor surfaces is an expensive nuisance because it makes people want to redecorate at shorter intervals than would otherwise be necessary. It is caused by any of four factors: temperature differences across the inner surface; local dampness; differences in surface permeability; and the amount of dirt in the contact air.

Temperature differences are influential because some dirt deposited is molecular, so the warmer the surface, the greater its molecular activity will be and the slower will be the deposition. The differences of temperature may be caused by differing rates of heat loss, like cold bridges, or differences of thermal capacity behind the surface. Dampness is a factor simply because dirt is then more adherent. Differences of permeability create staining because air which moves in and out of the surface leaves dirt behind as it goes in but does not remove it on departure. It is very noticeable on indoor fabric stretched over battens; there should always be a uniform backing to stop air percolation. The fourth cause, dirt in the contact air, becomes a factor when warm or cold currents swirl across surfaces differentially – for example, above radiators or when cold air infiltrates through a crack.

Internal cracking

Surface cracking indoors is another economic nuisance usually resulting from moisture or thermal movement of the inner leaf material. It mainly shows itself where the leaf abuts a main structural frame. If differential drying shrinkage wants to take place, nothing will stop it.

When brick or blockwork inner leaves are used with reinforced concrete framing, a horizontal crack sometimes appears along the central part of an inner leaf two or three courses above the floor in brickwork or one course up in blockwork. The cause is likely to be plastic deflection of the perimeter beam allowing one or two courses at the bottom of the wall to drop in the middle.

4.8.8 Firmness in construction

In my opinion the use of thermal insulating blockwork for the inner leaf, while offering improved insulation, has the disadvantage that its customary shrinkage has led to the loss of firmness in the holding of door and window frames which has been characteristic of much post-war construction. A return to a stronger, higher-density and lower-shrinkage inner leaf could correct this, but, of course, success then depends on having adequate insulation in the cavity.

4.9   Summary of the principal arguments

Five types of external cladding have been discussed in this chapter in diminishing order of weight: brickwork; pre-cast concrete; thin stone; rendering and thin sheet finishes, and a variety of inner leaves. In the everyday literature of the architectural and construction worlds these are usually discussed as if they were distinct kinds of building systems, but that is a mistaken approach which prevents perception of the fundamental fact that they are simply a family of ways of enclosing a cavity containing most of the required thermal insulation.

By making the cavity itself the first focus of discussion it became easier to think logically about moisture entry, types of jointing, ventilation of the cavity, the efficiency of the insulation, the differing roles of thermal mass in the outer and inner leaves, fire in the cavity, the management of wind pressures, and the differing needs for air and vapour permeability in the inner leaf, and this should be a useful approach in practice. At a secondary level, it should also help to clarify thinking about damp-coursing and about door and window openings in cavity systems (although doors and windows have a later chapter to themselves).

More ambitiously, perhaps, it may lead to computer software that gives more comprehensive and more accurate coverage of factors that have a bearing on thermal comfort and energy conservation, although some factors will need a better database than is presently available. In general too little use has been made of the link between the thermal mass of the inner leaf and the inertia of the heating system, and this applies with particular force to buildings in which stable temperature and humidity are needed.

References

1   BRE Information Paper 12/90, 'Corrosion of steel wall ties'.
2   Newman, A. J., 'Water penetration through walls', Municipal Building Surveyor Annual Conference & Symposium on Construction Problems, 1983.
3   Brand, R. G., 'High humidity buildings in cold climates: a case history'.
4   Allen, W. A., and Mills, E. D., 'Materials and techniques', *J. RIBA*, **61**, No. 8. 302–27, 1954.
5   Chin, I. R., Stecich, J. P. and Erlin, B. (of Wiss, Janney Elstrer Associates, Chicago), 'The design of thin stone veneers on buildings', *Building Stone Magazine*, May/June 1986.
6   Fitzmaurice, R., *Principles of Modern Building*, HMSO, 1938, p. 266.
7   Anderson, J. M. and Gill, J. R., *Rainscreen Cladding*, a Report for CIRIA, Butterworth, 1988.
8   Birkeland, O., *Curtain Walls*, Handbook 11 B of the Norwegian Building Research Institute, Oslo, 1962.
9   Garden, G. K., *Rain Penetration and its Control*, CBD40, National Research Council of Canada, Division of Building Research, Ottawa, 1962.

Further reading

1 Smith, P., 'Is the cavity a sacred space?' *Building Design*, 26 February, 1993. A criticism of the Building Regulations on Energy Conservation; the lack of allowances for trade-offs.
2 Ashurst, J., *Mortars, Plasters and Renders in Conservation*, Ecclesiastical Architects and Surveyors Association. A valuable reference for conservation work.
3 Bowler, G., 'Mortar enemies', *Building Design*, 25 September, 1992.
4 Hammett, M., *A Basic Guide to Brickwork Mortars*, Brick Development Association, Technical File, No. 23, October 1988.
5 BRE Digest 362, Mortars.
6 Spiegelkalter, F., Guide to the Design of Cavity Barriers & Fire Stops, BRE/CP7/77. Usefully informative though written in relation to the 1976 Building Regulations, now superseded.
7 Morton, J., 'Designing for movement in brickwork', Brick Development Association Design Note, July 1986.
8 Sims, I., 'Quality and durability of stone for conservation', *Quarterly Journal of Engineering Geology*, **24**, 67–73, 1991.
9 Gere, A. S., Design Considerations for Using Stone Veneer on High Rise Buildings, American Society for Testing Materials (ASTM).
10 BRE has published a series of Information Papers dealing with the choice of stones for concrete aggregate, conserving the hardest for the most demanding applications.
11 'Failure of the marble cladding on the Amoco Building in Chicago', *Building*, 2 September, 1988 and 28 April, 1989.
12 Davies, H., BRE Information Paper No. 7. Deals with the effectiveness of surface coatings in reducing carbonation.
13 Beckett, D., 'The influence of carbonation and chlorides on concrete durability', *Concrete*, February 1983.
14 BRE Digests Nos 325, 326 on Concrete, 1987.
15 BRE Digest 330, Alkali Aggregate Reactions in Concrete, March 1988.
16 Josey, B., 'GRP Cladding; Specification', *EMap Architecture*, 1994.
17 BRE Good Building Guide, Choosing Between Cavity, Internal and External Wall Insulation, 1990.
18 *Effects of Acid Deposition on Buildings & Building Materials in the United Kingdom*, Building Effectiveness Review Group report, HMSO, 1989.
19 BS 7543: A Guide to the Durability of Buildings, Building Elements, Products and Components, BSI, 1992.
20 BRE Digest 280, Cleaning External Surfaces of Buildings, December 1983.
21 BS 6270: Code of Practice for Cleaning and Surface Repair of Buildings, Part 1, 1982; Natural and Artificial Stone, and Clay and Calcium Silicate Bricks, Part. 2, 1985, Concrete.
22 Harding, J. R., Smith, R. A. and Brown, R. G. D., *Cleaning of Brickwork*, Brick Development Association, Note No. 2.

5 Applied finishes: tiling, mosaic, rendering and stone

Tiling, mosaic, renderings and stone are finishes typically applied to backings of brickwork, blockwork and concrete, both pre-cast and *in situ*. All can be successful but are not always so; much depends on interactions between the finishes, beddings and backings. We will begin with the backings and work our way outward.

5.1 Brickwork backings

5.1.1 Sulphate action

Bricks for backings will usually be commons – flettons are a familiar example in the UK – and quite a high proportion of these, probably more than half of all commons made in the UK, are likely to contain one or more of the sulphates of calcium, magnesium, potassium or sodium in potentially troublesome amounts. If trouble develops the consequences can be unpleasant, sometimes disastrous.

These sulphates usually originate in the clay itself but can come from other sources, such as sulphur gases in the kiln. They are partly decomposed or expelled during firing, but common bricks are relatively cheap, either because they are not fired to very high temperatures and may not therefore dispose of the sulphates effectively or because, as with flettons, the clay itself has a combustible oil content which contributes to the reduction of firing costs. Also, kiln temperatures are not always uniform and some bricks get less well burnt than others.

The key point is that all four of these sulphates are water soluble, and if brickwork containing any of them at high enough levels becomes damp for long periods, the salts concerned will slowly go into solution and react with a constituent of Portland cement, tricalcium aluminate, in mortar or render to form a strongly expansive crystalline substance, ettringite, at all interfaces.

This acts rather arthritically on brickwork jointing, expanding the brickwork vertically and horizontally and this, together with the formation of ettringite between the brickwork and the first-applied undercoat of a rendering, will cause detachment. The detached area usually falls (Fig 5.1), but if it does not and the brickwork behind it remains damp, this will continue to deteriorate. If freezing weather occurs, the degradation accelerates. Rendered garden walls of common brickwork very often demonstrate sulphate damage (Figure 5.2). Any brickwork damaged in this way has usually to be replaced, although sometimes it can be left to dry out behind a battened-out screen that lets air but not water get at it.

Figure 5.1
A fall of tiles forming the facework to a wall of keyed commons containing excessive sulphate.

Figure 5.2
A rendered garden wall of sulphate-loaded commons unprotected by DPCs.

The expansion is powerful and it is seldom practicable to contain it. In one case, four storeys of 112 mm fletton brickwork with a tiled finish on a rendered base lifted a concrete roof about 20 mm. Other cases have been seen where the movement dislocated and distorted window and door openings (Figures 5.3 and 5.4).

The sideways expansion is not as rapid as the vertical because there are fewer joints to expand, perpends are less tightly filled, and the interlocking of the bricks provides some restraint. But it happens (Figure 5.5).

The fiercest and quickest of the sulphates in reaction is magnesium, but calcium is the most widely found. Given time, all cause the same reaction to some degree if the brickwork stays damp. If the brickwork is kept dry or if it is exposed and can dry to normal atmospheric humidities, the trouble seldom develops significantly, sometimes not at all. Unfortunately, as explained below, tiling and mosaic and even stone finishes will all admit water but prevent drying, and they will all trap it into the brickwork if this gets wet during construction.

This is obviously a common occurrence, but the amount of water that can be held is not widely appreciated. Most common bricks are absorbent; and one or two types have been seen which had such active suction that they de-watered the contact face of a rendering so completely that the cement could not set. The unbonded rendering eventually fell away.

Impermeable paints applied to damp, high-sulphate brickwork can also prevent drying sufficiently to cause some trouble, but the paint finish itself usually degrades too; small blisters form, often with soft, furry crystalline material around them. A good rain can usually wash the lot away and the brickwork may be saved until it is repainted with a safer paint.

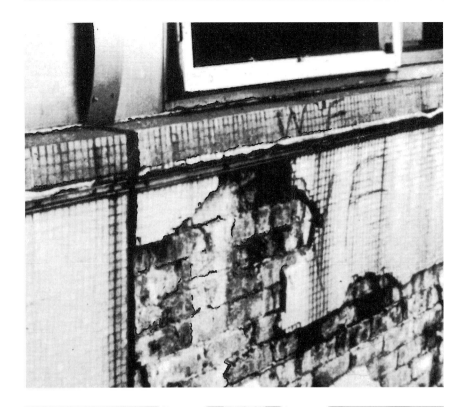

Figure 5.3
Sulphate expansion of common brickwork faced by mosaic on render. The upward thrust is shown by the curved cover strip. Severe damage to windows and walls. Both were replaced in a large programme of works.

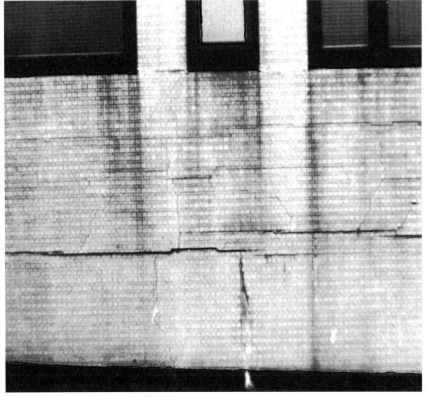

Figure 5.4
The signs of sulphate expansion. The facing is mosaic tiling on common bricks.

Sulphate risk levels

Although specification of acceptable limits of sulphate content is not a generally practicable way of avoiding trouble, it may be useful to know that amounts of individual sulphates greater than the following can be regarded as having a degree of risk, as will also be the case if the total of any combination exceeds about 0.5 per cent:

Calcium 0.5   Magnesium 0.15–0.25
Potassium 0.03   Sodium 0.03
Quantities over 0.1 per cent all call for caution. The percentages are by weight of a dry brick.

Unfortunately the current British Standard[1] offers no numerical guidance about risk percentages of soluble salts, only that efflorescence should be no worse than moderate, which is not much help to architects and specifiers who must not take chances. I have therefore derived the figures shown from some quoted long ago by Fitzmaurice.[2]

Relevant figures for specific brick types are seldom, if ever, quoted in trade literature, but manufacturers should all know what they are for their own products and may indicate a knowledge that they could be at risk levels by recommending the use of sulphate-resisting cement (SRC) for external brickwork to which a finish is to be applied. This cement avoids or minimizes the reaction by having most of the tricalcium aluminate removed in manufacture. The cost is higher but a reasonable safeguard results.

Manufacturers who know that their products are free of sulphates at risk levels will doubtless be willing to produce a suitable form of warranty on request, and this should make the use of SRC unnecessary. But if SRC is recommended by a brickmaker, it should be regarded as essential not only for the soundness of the work but also because of liability risks. It must be used both for the mortar and for at least the first undercoat of any applied finish, but preferably for all coats. It would not seem inappropriate if makers of bricks containing significant levels of sulphates were required to state percentages in their trade literature.

It should be a matter for concern that the risks of using commons with potentially excessive sulphate content has been known and given publicity at intervals ever since research at the Building Research Station clarified the nature of the problem at the end of the 1920s, yet it seems that the collective memory of the building industry has been unable to take in and hold the knowledge. It still comes as a surprise to some architects, builders, clerks of works and sometimes even building control officers when sulphate trouble occurs.

Case notes

Some high-rise blocks of flats, built under a design/build contract, had an outer leaf of fletton brickwork with undercut grooves, finished with an applied mosaic. The DPC was black polythene but amazingly the DPC joint had been pointed and sealed by mastic (Figures 5.5 and 5.6). Continuous runs of timber windows were set on a simple Z-shaped aluminium extrusion to form a cheap weathercast cill (Figure 5.7).

What happened was surprising. Floor deflection squeezed the windows down and rocked the cills back over the narrow strip of bedding mortar on which they sat so that they acted as a gutter instead of a weathercast. The water ran through their end-to-end joints opened by thermal movement, and wetted the brickwork behind the mosaic. It also collected on the DPC because the mastic seal in the DPC joint prevented its escape. It got through unsealed joints in the polythene and ran out on to the floors inside. The brickwork expanded sideways by sulphate action using the DPC as a slip plane and some mosaic areas fell off, demonstrating that even with rendering bedded into undercut grooves on bricks the mosaic could be detached by ettringite.

Figure 5.5
Corner damage to a high-rise block of flats by sulphate expansion sideways. The brickwork slid along a polythene DPC hidden by the unwise mastic pointing.

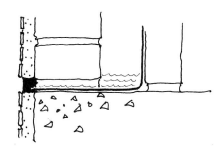

Figure 5.6

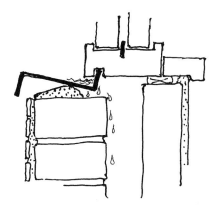

Figure 5.7

Sulphates can also attack metals. On one building examined, the galvanized wall ties were largely eaten away in eight or ten years and the integrity of the walls was destroyed.

Trouble with sulphates can be a disaster for all concerned. The key to its avoidance is to be able to recognize situations where construction dampness will be trapped into common brickwork by the finishes applied to them, or where the finishes themselves may draw it in afterwards. Some of these risk situations are discussed at later intervals in the text.

5.2 Concrete blockwork as a backing

Concrete blocks are normally either dense or aerated. The most suitable for backings are dense blocks with an open, rough surface which provides a good key for the application of the prime undercoat of the bedding render. There will be some modest initial shrinkage of such blockwork and then it should settle down to a fairly stable state. Joints should be raked to maximize the key for any render.

Aerated concrete blockwork has a much higher initial shrinkage and a more absorbent surface, often sufficient to de-water the contact face of the first render undercoat and diminish the quality of the bond. The shrinkage can then cause detachment of the rendering. See also Section 5.4.4.

5.3 *In-situ* concrete as a backing

5.3.1 Drying distortion

There has already been mention of the way reinforced concrete framing shrinks and sags as it dries during its first few years. Beams deflect between columns so that they form a shallow sinusoidal curve, and floor and roof panels deflect similarly. Columns shrink, and those on the perimeter, being loaded eccentrically, also become slightly sinusoidal vertically. It augurs ill for finishes applied direct (Figure 5.8).

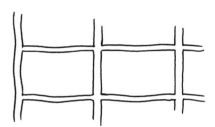

Figure 5.8
Typical concrete frame deflections and column distortions.

Case note

In the worst case of this I have seen the floors of a multi-storey building were cast as thick slabs in 50-ft spans without beams in order to save height, and the floors were carried by a perimeter ring beam bracketed in from exoskeletal columns. The ring beam came up to cill height to carry continuous window runs.

The floors sagged severely and frustrated the intention to use prefabricated partitioning, the ring beam deflected and had a twist at mid-span positions caused by the floor sag, and the exoskeletal columns became twisted by the eccentric floor loads bracketed inward from them. The window system and some applied finishes suffered badly.

When beams are very deep they may be cast in two pours. The reinforcement crosses the interface well enough apparently for the strength of the beam to develop, but it has been found that the bond may nevertheless not be true. Microslippage and differential shrinkage probably explain this. A disturbance line may show through mosaic, and tiles are sometimes dislodged along it if they have been applied direct to the concrete.

As previously noted, steel framing with concrete cover behaves differently, for the steel does not shrink nor does it flow plastically into some other form, and the concrete then behaves like the steel.

5.4   The finishes

With all these foregoing remarks about backings in mind, consider what the finishes may do to what is behind them, and what this in turn may do to the finishes.

5.4.1   Undercoats

Tiling and mosaic are applied over mortar undercoats built up like those described in Section 4.5 for rendered finishes and only a few further comments are needed. These concern the reliability of adhesion, first, of the primary undercoat to the backing, and then of the successive coats to one another.

As was made clear in Section 4.5, stainless steel mesh or expanded lath is by far the best base on which to build up a mortar backing. It gives an absolutely positive key, and if the mesh is continuous over its ribs it provides reliable reinforcement restraint against shrinkage. Highly textured brickwork with raked joints is quite a good backing, but smooth brickwork or concrete is not to be trusted (Figure 5.9), even with so-called

Figure 5.9
The remains of a tile collapse on smooth concrete.

sparrow-pecking. Roughening of the whole surface is better but not suffi-ciently reliable for multi-storey buildings where falls of detached finishes can be lethal. Undercut grooves made by shaped rubber form-liners give reliable attachment.

Good intercoat bonding as described in Section 4.5 is absolutely essen-tial, and this underlines the fact that one is totally dependent on intelli-gent and careful workmanship for the reliability of finishes applied to rendered bases, not always found on sites.

5.4.2   Tiling

Tiles are made of burnt clay. They therefore come dead dry from the kiln and begin to expand as they encounter and absorb atmospheric moisture. When tiles are used as an external finish, rain works its way through the joints to permeate the bedding so that they continue to be fed by dampness and their expansion becomes maximized over a few years.

Unfortunately it has been customary to recommend dense jointing mortar for tiling[3] with the result that expansion becomes collective across large areas. This has led to another recommendation, to place a soft joint through the tiling and its bedding at intervals of several metres to accommodate the expansion.

This is misconceived. Although such tiling must expand, its mortar bedding will be trying to shrink, and the effect of the soft joint is simply

(a)

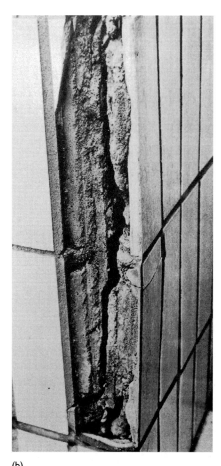

(b)

Figure 5.10
The column has shrunk and the tiling has expanded, forcing the bow-legged detachment.

to remove resistance to the tiling expansion and allow shearing detachment to take place between it and the shrinking bedding near the soft joints where the movement differentials maximize. Hollow areas indicating detachment of tiling are commonly found near the soft joints.

The location of the shear plane depends on where the adhesion fails. It may be between the tiling and its immediate bedding or between the bedding and the next undercoat, but more often between undercoats themselves or between the prime undercoat and the backing. If the tiles are pressed and have no undercut key on the back, they will probably detach readily from their immediate backing, but this type should never be used for external work. If they are extruded and have undercut grooves on the back, then they usually take their bedding with them when they come away, and any potential shear plane moves farther in.

Case note

A pair of modern buildings built in the mid-1960s provided some further evidence about these matters. In one case the concrete was cast against rubber form liners providing undercut grooving, and the tiles had undercut grooves on their backs. The pointing mortar was not strong. Thirty-five years later the tiling was still soundly in place. On another building finished a couple of years later, the concrete was smooth and the tiles were pressed and therefore without such grooves. These detached themselves over large areas. In both cases a lot of water ran down behind the tiling and entered the building.

The amount and rate of tile expansion depends on the absorption coefficient of the tiles. Differences as great as 0.2 per cent to over 4 per cent have been found from different manufacturers, tested over the rather short period required for standard tests. (The *in-situ* expansion will be greater when the tiling is damp for prolonged periods.) If external tiling is to be used in a damp climate it should be done with tiles with very low expansion coefficients, e.g. of the order allowed for lining swimming pools.

Dampness behind tiling requires explanation, for it is not a wholly water-shedding finish.

Most tiling is done by applying a pat of mortar to the back of a tile and tapping it into place. The tiler judges how big a pat is needed in order to get enough coverage to hold the tile without wasting mortar, and this results in lack of filling at the corners of tiles and sometimes along parts of the sides and ends (Figure 5.11).

Figure 5.11
Some tiles that had come away from concrete.

When the pointing is done for the joints, it has to be made against pressure if adequate contact to the sides of tiles is to be made, but the lack of bedding behind some parts of the joints prevents consistent pointing pressure along them. The tile-to-mortar bond in the jointing is therefore uncertain and is made more so if glaze has run over the edges of tiles or if they are dirty or dusty before pointing, as often happens. Also, inexplicably, some tiles have edges shaped in ways that actually prevent good filling of joints (Figure 5.12).

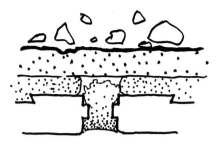

Figure 5.12

These poor bonds soon get loosened by the temperature changes to which external finishes are subject and capillaries or bigger gaps develop along the tile edges, their effectiveness maximized by the glaze on the tile edges and the dense mortar. Every square metre of brick-size tiling has some 30 m of joint-to-tile interfacing and probably 10–20 per cent of it ends up as capillaries or larger gaps, drawing in rain or allowing it to be pushed in by wind.

It is also likely that thermal pumping will be at work. When tiling is done as described, the voids left behind the tiles at corner meeting points and along their edges create the right conditions for thermal pumping, and in cold and rainy weather it is to be expected that water will sometimes be drawn in when the finish and bedding is chilled and sub-atmospheric pressures will develop in the voids. Free water is often found behind tiling in investigations and evidence for it is shown in Figures 5.14 and 5.15. On some buildings examined, the backings had evidently become saturated so that heating of the tiling by sunshine caused expansion of the contained water and made it emerge from joints. In one case this forced some of the mastic out of the soft joints (Figure 5.13).

On a few occasions attempts have been made to re-attach loose tiling by injections of epoxide to fill the voids, the interface gaps, and interstices between bedding coats. It seems that the epoxide itself gradually absorbs some of the moisture behind the tiling and expands. This may be innocuous when it is only a thin layer filling a thin interface gap, but

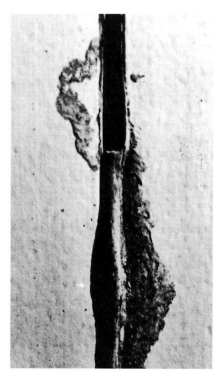

Figure 5.13
Soft-joint mastic forced out of its joint by solar heating of water behind the tiling.

Figure 5.14
Plenty of water gets in behind tiling as shown by the runnels behind the joints.

Figure 5.15
Water behind and beneath tiling squeezed out by solar warming and expansion.

in the greater volumes of the voids at tile corners and along under tile edges, expansion amounts have been large enough sometimes to bend the tiles, causing them to crack and gradually break up.

Clearly not only will the bedding usually become damp behind tiling but the backing will gradually dampen too, and if this happens to be of oversulphated common brickwork, the conditions will exist for ettringite to develop and cause expansion and detachment, as described earlier.

Tiling on concrete blockwork

Dense concrete blockwork can provide a good mechanical key for tiling, as we said in Section 5.2, but aerated blockwork is not likely to be satisfactory as a backing because of its shrinkage.

Tiling on in-situ concrete

When tiling is applied to new concrete, conflicts are inherent. The concrete is shrinking, columns are becoming shorter and beams are sagging as it dries, while the tiles are taking up moisture – some of it from the concrete – and expanding. The tile jointing mortar is usually strong, so that the expanding tiles turn the tiling into rigid plates or casings trying to get larger, while behind them the concrete is shrinking and deforming.

The severity of the conflict depends on several factors, chiefly the concrete mix, the coefficient of moisture expansion for the tiles and the density of the jointing mortar, but reliable adhesion is clearly unlikely and it may be prejudiced by one other factor. Tiling is a compelling discipline if the appearance is to be right. The jointing must be consistent, the tiles themselves must be whole, not cut, and the final surface must be a true plane over the whole facade. But they are being applied to carcasing concrete which it is not economic to build accurately, so the bedding mortar often has to be built up to considerable thicknesses and widths to provide a true tiling base. Thicknesses of 35 or 40 mm have often been found on bad carcasing, and sometimes the weight is such that the coats pull away from one another before a proper bonding set has taken place and leaves them vulnerable. The next technique to be described is more reliable.

Tiling on metal lath

Metal lathing can be framed on and around concrete carcasing to an accuracy sufficient to avoid the need for any thick build-up of bedding mortar, and the mortar itself will have far more reliable adhesion than it can ever get to concrete. Its own shrinkage will also be restrained, so that its reliability as a tiling base will be improved, and the system as a whole will have reasonable freedom from most of the shrinkage and distortion of the concrete carcass.

As already noted, the lath must be of stainless steel and be continuous across any ribbing so that it provides uniform restraint, but given these improvements, plus the use of tiles with low moisture expansion coefficients and undercut backs, reasonable reliability should result.

Examples seen have behaved well. Soft joints had been incorporated at intervals of 3 m but no evidence of adjacent detachment due to movement differentials was found.

Need for research

No research appears to have been devoted to tiling detachment risks and therefore it is not possible to say whether the soft joints are necessary when the backing is metal lath, what limits of moisture absorption and expansion are acceptable, or what increase in risk there is if darker coloured tiling with its greater thermal movement is used in place of the white or the light colours most commonly used in the UK.

Tile glazes

The glaze on tiles is an adjusted clay slip, such that it emerges in compression from its firing. The purpose of this is to allow the body of a tile to undergo some moisture expansion without cracking the glaze; the compression merely relaxes. This explains why most bathroom tiling stays crack-free. If the glaze on tiles cracks *in situ* it may mean that tile expansion has been greater than expected and has put the glaze into tension. This may signify excessively damp backing or tiles with a high moisture expansion coefficient, or simply badly made tiles. One should establish which it is.

Designers familiar with the ancient traditions of colourful tiling in the Middle East may ask why it should apparently work so well there and its success be so dubious in Britain (Figure 5.16). In the Middle East, the materials were different, the climate drier and, so the author was told in Isfahan, buildings were usually allowed several years in which to settle down before being tiled. Nevertheless, the tiling is said to need quite a lot of maintenance.

Figure 5.16
Tile finishes in Isfahan.

5.4.3 Mosaic

Mosaic is a very ancient technology. It comprises small pieces of white or coloured stone, ceramic or glass, known as tesserae. They are normally pre-fixed in flat colour or patterns to paper with a water-soluble adhesive, and the resulting sheets are then applied paper-side out to a thin bed of cement-rich grout on a rendered base. When this has set, the paper is removed and the joints are filled by further grout applied by squeegee, and the excess is wiped off.

Because grout is either neat cement and water, or at any rate very rich in cement, it will shrink as it sets and dries. The jointing between tesserae adds up to many metres per square metre and not much of it will be contact-perfect. Water gets under the edges of the pieces and in frosty weather it can freeze and exert leverage, gradually loosening some of them. Water also penetrates microcracks in the bedding grout or diffuses through it because of its thinness and then gradually permeates the undercoat build-up and sometimes also the backing. As with tiling, the water that gets into the undercoating and backing will be slow to dry out, but unlike tiling there is no collective expansion of the finish; it is dimensionally stable.

Again, as with tiling, the top undercoat of the backing must be brought consistently to the correct surface alignment before mosaic is applied.

Application to brickwork and blockwork

Since mosaic allows moisture to get through to the backing, its application to oversulphated common brickwork will lead to the latter's expansion unless sulphate-resisting cement is used. Sulphate expansion behind mosaic typically produces horizontal cracks every five or six brick courses, connected by uneven vertical cracks at intervals of a metre or so (see Figure 5.4). Disbonding becomes general. Window and door frames are likely to get distorted and jam (see Figure 5.3).

Case note

A case occurred in a wet part of Britain where mosaic on a render base had been applied to heavily sulphated common brickwork on a multi-storey building. Much distortion of the brickwork took place (see Figure 5.3), and when freezing occurred after a period of rain one patch of mosaic and bedding 21 m² in area detached itself and fell.

Dense concrete blockwork that has low moisture movement and a good mechanical key for the bedding is unlikely to give trouble to mosaic but if blockwork shrinkage is high, detachment is likely.

Mosaic should not be applied across joints where movement can occur for any reason, for a disturbance line will develop. This will occur, for example, if mosaic is applied across junctions between wall panels and the structural frame, or across the pour-line on a deep concrete beam done as a double pour, and it frequently disfigures an otherwise pristine appearance when applied to large areas of concrete in which cracking due to shrinkage or deflection has not been controlled successfully by reinforcement.

Mosaic applied to rendering on metal lath as described for tiling should be even more reliable than the tiling in view of its freedom from moisture expansion.

5.4.4   Rendering as a self-finish

Rendered finishes have already been discussed in Sections 4.5 and 5.4.1 and need few further comments in the present context. When properly made and applied, a rendering keeps water out very well but if applied to oversulphated common brickwork that has become wet during construction, it can sufficiently prevent drying to allow ettringite to form and then expansion and cracking will result and allow progressive wetting and deterioration to develop.

When applied to dense concrete blockwork it should generally prove reliable. On aerated blockwork it is a different matter.

Case note

Several houses were built in a very wet area using aerated concrete blockwork for both the inner and outer leaves of the cavity walls, and a strong rendering was applied as a finish. The manufacturer of the blocks advised on the location of soft joints through the finish and the outer leaf to control crack location.

The houses were centrally heated, the blockwork dried out, and cracking occurred in the rendering at numerous other places as well as at the intended soft joints. Accidental damage at one of the outer corners of one of the houses then unexpectedly revealed that there was a gap of some 8 or 10 mm between the rendering and the blockwork backing at some positions. The rendering was found to be virtually free-standing, stabilized chiefly by such returns as occurred at window and door reveals.

5.4.5   Ancient renderings and modern paints

An unexpected discovery was made about old renderings in the course of analysing a failure.

Figure 5.17

Case note

This was on a sixteenth-century mansion with traditional thick walls of rough stone and soft mortar, rendered and with quoins of dressed stone. As usual there were no DPCs. This was the original rendering, a traditional mix of lime, dung and sand as found in many parts of the world, traditionally decorated by lime-based colour washes. It was all in excellent condition despite its 400 years.

The owner had been persuaded to apply a modern paint in the expectation of getting much longer periods between repaintings. Instead he found that after the first winter the paint developed huge, brittle blisters behind which the rendering was wet, covered by green algae, and coming away from the wall. It could not be rescued (Figure 5.17).

The investigation revealed interesting complexities. Unknown to the progenitors of the traditional mixes, dungs are polymucosaccharides. Saccharides are sweeteners, acidic, and in a mix with lime, which is an alkali, they react quickly to form an acid/alkali cross-linked molecular lattice, having what is technically known as the structure of a gel. This reasonably explains why such mixes have a quicker set than might otherwise be expected, and also greater strength, although the lime would gradually have hardened anyway by the absorption of atmospheric carbon dioxide.

So how did the paint come to cause so much mischief? In these old walls the ground moisture wicked its way up through the core, much of it transpiring harmlessly to atmosphere through the porous rendering and colour washes. However, at the time the paint in question was applied, most paints had gone over to alkyd resin as a base which not only made them far less permeable to water vapour than the traditional colour washes but also made them unstable in contact with lime. Decomposition of the paint film then took place by hydrolysis, causing expansion and embrittlement. The expansion and impermeability combined with the vapour pressure trapped behind created the blisters, which soon became friable and broke, revealing in this case the wet, green, and weakened rendering.

The reason for the weakness is probably this. The rendering is wet because the trapped vapour condenses in it, behind the paint. The mucus gel of the polymucosaccharide then swells like any gel in water, and the forces involved in the swelling weaken the polymer lattice. Algae would grow because the alkyd resin would exclude oxygen and this would favour its growth in the prevailing dampness.

In the case examined, the 400-year life of the render was terminated in a matter of months. Microporosity has been restored to most alkyd resin-based paints but they should be evaluated nevertheless for safe use where permeability matters. And it is a reminder, which experienced conservation architects and surveyors will not need, that traditional practices often have evolved a physical and chemical balance that it is unwise to disturb.

5.4.6   Stone as a casing or applied finish

Stone veneers typically 100 or 120 mm thick have sometimes been applied direct to oversulphated common brickwork. The back of the stone or the face of the brickwork may be coated with a paint to prevent sulphate-bearing moisture from getting out to stain the stone, but this will not prevent rain from working its way inwards through stone joints. Expansion of the brickwork then follows unless sulphate-resisting cement has been used in its mortar. The process typically develops slowly because stone joints are narrow and the stone slabs are large, so that access for moisture is restricted. When the expansion of the brickwork begins to force the stones apart, moisture finds it increasingly easy to enter and progress becomes quicker (see Figure 5.18).

Stone has occasionally been used as formwork for large and thick solid concrete walls. Coventry Cathedral is an example, and a question that was debated was whether to paint the back of the stone to prevent possibly damaging adhesion as the concrete shrank or to leave it unpainted and risk some staining from the concrete. In the event it was painted, but as a matter of interest it was noted that with the 4-ft concrete thickness to be cast, its drying would take some 20 years to reach equilibrium. Without the paint it would take place in perhaps half that time by being able to dry to both sides.

Figure 5.18
A stone stair parapet broken up by sulphate expansion of the brick core.

A postscript

Figure 5.19
Not all building materials are manufactured to British or any other standards. This brickwork was suspect and the author found the bricks so bad that he could put his pocket knife in. At one penetration some strange mucous-like material came out. You have been warned.

References

1   BS 3921: 1985 (1995); Specification for Clay Bricks.
2   Fitzmaurice, R., *Principles of Modern Building*, HMSO, 1938.
3   BS/CP212, Part 2, June 1966, dealing with external tiling. Code of Practice for the Design and Installation of External Ceramic Wall Tiling and Mosaics (including terracotta and faience tiles). Republished as BS 5385, Part 2, in February 1978, and again in 1991.

Further reading

1   BRE Digest 362, Building Mortar, 1991.

6 Curtain walls

6.1 Historical background

The concept of the fully glazed wall developed fitfully but persistently over some 500 years and was presumably therefore an architectural idea only awaiting the necessary technology to become a reality. Surprising strides towards this goal were made in the final stages of the Gothic era, when buildings such as Bath Abbey and King's College Chapel in Britain and La Sainte Chapelle in Paris took windows to the point of forming virtually all of the wall that was not needed for roof support. To my mind the architects concerned, and their patrons, must have been urged on by a vision of ultimate purity of form and structure in the Gothic idiom as they pushed it to these limits at the end of its long evolution.

The idea had something of a counterpart among houses in the Tudor secular world, for one can see increasingly large areas of glass brought into their design during the sixteenth century. Their architects doubtless enjoyed exploring its potential, very much as we would have done, but their patrons, the great magnates, had other motivations. It was an age when wealth, if you had it, was flaunted. Glass was expensive, so the more you could display of it, the richer you could be seen to be, and the higher your social status. Bess of Hardwick was the role model and her last house the ultimate example of its time.

Classical formats of design which took over in the Renaissance precluded this form of display and large glazed areas passed from the

Figure 6.1
Hardwick Hall, Derbyshire, the last home of Bess of Hardwick, Countess of Shrewsbury and embroidery friend of Mary Queen of Scots. A notable 17th-century example of wealth and self-esteem displayed by the use of glass. Photo: David Medd.

architectural vocabulary until conservatories began to be needed for gardens in the eighteenth century. Initially these were framed in wood, but with the realization early in the nineteenth century of the potential of combining iron and glass, not only conservatories but also railway and other industrial architecture found enclosures of this kind logical and exciting, culminating exuberantly in Paxton's Crystal Palace of 1851.

It was an infectious building and large glazed walls and roofs came into existence on several European and American buildings. Wanamaker's Department Store of 1859 in New York was a leader, and by the turn of the century the idea was becoming one of the stigmata of the Modern Movement. A big step forward was taken by Shreve, Lamb and Harmon in their Empire State Building of 1929 in New York where they introduced aluminium spandrels, and a year later Holabird & Root finished a building for the A. O. Smith Corporation of Milwaukee which used walls largely of glass and heavy aluminium extrusions. In Britain, William Crabtree's building for Peter Jones in Sloane Square, London, in the late 1930s had notably elegant curtain walls.

After 1945 a major advance was made by Pietro Beluschi in his Equitable Building in Portland, Oregon. This was the first large building to be totally sheathed in glass and aluminium, the first to be fully air conditioned and double-glazed, and the first to have a travelling crane for window washing. At about the same time came the Lever Building in New York by Gordon Bunshaft of Skidmore, Owings and Merrill, soon to be followed by Harrison and Abramovitz's building for the United Nations and the Seagram Building by Mies van der Rohe and Phillip Johnson.

Despite this robust post-war history, problems have been experienced and the learning curve in overcoming them has not been at all smooth. Aluminium has become the customary metal for framing but its high coefficient of thermal movement has contributed significantly to leakage and other troubles. Various ways of accommodating glazing reliably in this thermally restless framing have been developed but not all have been successful. Requirements for higher and higher thermal insulation have created complications, too, and the fact that curtain walling can become boring has given rise to a succession of design innovations that have often created new learning problems. The ingenious simplicities of the Crystal Palace are behind us.

In fact, the complexities of reliable curtain wall design are now such that it is generally wise to work with proprietary systems which have undergone and survived rigorous testing in the course of development, because modern test rigs are large enough to reveal problems of thermal movement in the framing and can subject trial areas to combinations of wind and rain sufficiently exacting to simulate conditions of exposure even on high buildings. Therefore what will mainly be discussed here are the technology variants which one is likely to encounter in proprietary systems, some of the reasons for them, and matters that require consideration in using them.

6.2 The technology

6.2.1 Thermal differentials

The see-through parts of curtain walls will now usually be sealed double-glazed units, and the opaques are sometimes sealed containers of some kind with a thermal insulation filling. Both the glazing and the containers will therefore have one side that maintains a temperature close to that of the indoor climate, while the other is subject to an outdoor range of temperatures. This causes bending due to different thermal movement on two connected faces, the outer trying to get larger or smaller than the inner (Figure 8.6, p. 166). This imposes a lot of stress upon their edge seals, and sometimes these fail. A proprietary technique for incorporating thin sheet stone on infilled panels has been developed by forming the panels as steel boxing containing the insulation and carrying the stone on hanger fixings on the face of the box.

6.2.2   Thermal insulation and condensation

A thermal break is usually interposed between the parts of mullions and transoms exposed to outdoor and indoor temperatures to avoid outward and inward bowing of the mullions due to thermal differentials and to remove the risk of condensate forming on the indoor surfaces in cold weather. It should be said, however, that if condensate forms on mullions or transoms unprotected by thermal breaks, the RH values indoors are above what they should be.

Mullion and transom framing is sometimes associated with cold-formed sheet metal to create larger casings, usually around structural elements of some kind or another and what can happen inside them is interesting.

Case note

A multi-storey office building was built in which the floor loads around the perimeter were carried by mild steel tension bars hung from the roof. The suspension steel was in casings which formed part of the curtain wall design. Problems arose which made it necessary to open some of the casings and these were found to be very damp inside, and with the suspension steel was in an advanced state of corrosion. This occurred only 15 years or so after construction and the building required very substantial remedials.

It was not possible in this case to distinguish between rain leakage and condensation as the source of the dampness; both were probable. Wetness in any inaccessible voids in a building envelope is not acceptable and casings of this kind should be ventilated. They should not contain insulation, for this will increase condensation and thermal movement.

6.2.3   Outdoor/indoor sound insulation

The sound insulation of curtain walls is largely determined by the sound reduction values of the glazing or the infill panels, whichever is the lower, and any air leakage in a system will reduce it. The most likely insulation will be in the range of 30–40 dBA. Where higher levels are required, consistently greater weight and rigidity will be needed and specialist acoustic advice should be obtained.

Some curtain wall systems can accommodate triple glazing by the addition of an internal opening light. Both sound and thermal insulation will benefit; but see Section 6.2.7 below.

6.2.4   Fire

No walls should be able to be ignited readily by any external source. However, it should be noted that aluminium loses its strength much more rapidly at temperatures generated by fires than does steel, for example, and can therefore collapse quicker, sometimes in a matter of minutes.[1]

Storey-to-storey firebreaks must be provided in the linkage to the primary structure. These are usually fixed to the edge of the floor, not to the more vulnerable curtain wall, and they may have to prevent storey-to-storey sound leakage as well. It is important to work out a reliable technique at an early stage in design in order to avoid the greater risks of failure that can easily accompany later improvization.

The appropriate spread of flame classification for materials on the external face of a curtain wall is Class O.[1]

6.2.5 Assembly

The framing

When mullion and transom framing is used, the tops of the mullions are usually fixed to the primary structure. Their thermal movement is then usually accommodated by a sleeve and spigot arrangement at top and toe, sometimes with a compression gasket at the junction to keep a seal. The same technique may be used to accommodate transom movements where they meet the mullions. The thermal movement is usually minimized by having the main body of the mullion and transom behind the glass line and behind any thermal break in the system.

The amounts of water getting into the mullion and transom system are minimized by having the glass units held by snugly fitting gaskets, which are clamped tight to the framing by strong cover strips screw-fixed to a shaped flange on the mullions and transoms (Figure 6.2). Such an arrangement is termed front-sealed and drainage of any leakage that may occur is provided behind the cover strip.

The size of the gap necessary to accommodate thermal movement in mullions depends on three factors: the length of the units, the ambient temperature at the time of fixing, and any creep or deflection to be expected in the primary frame of the building if this is of reinforced concrete. So far as ambient temperature is concerned, what has to be allowed for is the maximum range and consequent movement expected to take place due to exposure before the building is enclosed.

As regards creep and deflection in a primary frame of reinforced concrete, an allowance must be made for column shrinkage at 1 mm/m but no allowance is likely to be needed for beam deflection if deflections on all storeys will be similar. The roof beam may deflect less because of its smaller loadings, and if the curtain wall goes down to solid construction at the bottom, there will be no deflection there.

Internal drainage

In some types of curtain wall where the assumption is made that total exclusion of rain is not practicable or is not worth the cost, the design of the outer frame elements is modified to provide for drainage of the anticipated positions. The air-tightness of the drainage space is ensured by a seal on its inner side to prevent rain being blown through, and cavities are formed in which wind pressure equalization will occur and avoid the inward carriage of rain (Figure 6.3).

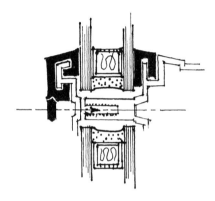

Figure 6.2
A gasket concept for front sealing.

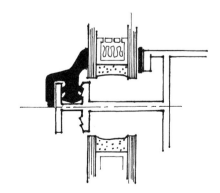

Figure 6.3
A gasket concept for the drainage arrangement.

6.2.6   Mastics and gaskets

Where a design requires an *in-situ* application of mastic, silicone is now usually preferred because of its strength and durable elasticity. Whatever type of mastic is used, it should be in an accessible position and should be inspected at reasonable intervals, probably of not more than five years, as part of planned maintenance.

The surfaces to which mastic is applied must be thoroughly clean and dry and application must be expert. It looks fool-proof and easy to apply but this is wholly misleading. Even when expertly applied to smooth surfaces of glass or metal, adhesion of more than 70–80 per cent is unlikely.

Gaskets are made of neoprene, EPDM or silicone, moulded accurately to shapes and sizes required, with vulcanized corners. They must fit snugly but should not be left overstretched *in situ*, nor should they clasp the glazing or panel infills overtightly or they will fatigue. They are precision-made.

Silicone fixings

'Structural silicone' is a term used to describe silicone moulded onto components and then used to fasten the component in place. It is proving to be a valuable addition to the technology. An example of its use is described below.

6.2.7   Triple glazing

Triple glazing is one way of meeting high standards of thermal and sound insulation. One example makes use of structural silicone to fix an outer layer of glass to the mullion-and-transom framing, while sub-frames fixed to the sides of the mullions and transoms carry conventional double-glazed lights which are hinged to open to enable the inner side of the external glass to be cleaned. The structural silicone is bonded to the back face of each pane of external glass around its edge, which carries a metal clip enabling it to clasp together the sub-framing and the mullions and transoms. The attachment protects the outer edges of the framing in such a way that no other thermal breaks are needed. No trim shows externally on such arrangements; there is merely a narrow gap between the panes. The interspace behind the outer glass needs to be vented to allow vapour pressure equalization with the external air to take place.

6.2.8   Mechanical fixings

Where a screwed or bolted fastening has to accommodate thermal movement, slotted holes with washers are necessary. The screws and bolts must be of types that will not unwind by vibration (section 4.7.2), and spring loading is likely to be needed to allow for slippage due to thermal movement. The washers must be large enough to cover the slot well throughout the whole range of possible thermal movement.

Slots are also necessary to allow for fixing tolerances because curtain walls are precision-made systems which have to be fitted to the less precise primary framing of buildings. The fixing bolts are tightened rigidly because subsequent thermal movement is accommodated in the curtain wall framing.

Mullion fixings have to be robust because they carry the dead load of the assembly. They must also be able to transmit wind pressures to the primary frame.

All parts of a curtain wall system must be able to withstand pulsating winds of the strengths to be expected at the height and geographical location of the building. The glazing and infill panels must do so without much bending. Their gasket seatings must transmit the forces to the transoms and mullions without significant slackness, and the transoms and mullions must be sufficiently strong and rigidly connected to avoid wind vibration of the system as a whole.

Sub-assemblies

Where construction time has to be minimized, curtain walling can be designed for factory assembly in sections to be lifted into position. In the industry this is termed 'unitized' construction. Special detailing may be needed where areas of wall have to be omitted temporarily to allow access for the contractor's use. Factory assembly has the advantage of providing better working conditions for precision work than are likely on site.

6.2.9 Chemical incompatibilities

There are several mechanisms by which chemical damage may be caused on curtain walls:

*   If certain dissimilar metals are used together there is a risk of corrosion even if the electrolyte is only water washings from one metal to the other. Contact is not necessary, although the reaction is then quicker. Contact between zinc, lead and aluminium should be specifically avoided. The same risk occurs when an inappropriate metal is used for screws, and this can easily be overlooked. PVC tapes and nylon washers can sometimes provide sufficient separation. See also the main discussion of corrosion in Chapter 10.
*   Alkaline overwashings will damage aluminium and galvanized or anodized coatings and must be avoided. Their principal sources on buildings are cements, lime and limestone. Protection against mortar droppings during construction is essential.
*   Neoprene and EPDM, two of the synthetics used for gaskets, can be damaged by petrol-based materials such as bituminous waterproofing liquids, felts, asphalt and by silicone in mastic form. Silicone in its turn can be damaged by petrol-based materials, acids, mineral oils, neoprene and toluene. If silicone is used, it is generally best to avoid the other two materials. As a general rule, always check compatibility between seals and mastics.
*   Silicone can be stained by contact with bitumens, and silicone oils can stain and contaminate some types of stone.
*   If wood products are incorporated into curtain walls they must be kept dry. When wood gets damp the dampness becomes acidic and can lead to metal corrosion either by contact or by acidic humidity. See the discussion of corrosion in Chapter 10.

6.2.10 Leakage mechanics

Curtain walls are effectively non-absorbent. In rain they are covered by a coat of water usually flowing downwards but sometimes cross-wise or upwards, depending on the wind forces on the face of the building. Any small gaps will have smooth, non-absorbent walls which offer maximum capillarity to water, and if assembly workmanship has not been good, leakage may occur right through the wall. Makers of proprietary systems have sometimes claimed that these are less likely to suffer than custom-made systems, but experience suggests that the risks are much the same.

Whether gaps will open up in service depends largely on the stiffness of the framing, glazing and panels, and on the accuracy of the gaskets and their seating. Wind forces will pulsate and attempt to bend individual panes of glass and panels inwards and outwards vigorously so that the gaskets will be subject to bending moments around the edges of the panels or the glazing and to alternating compression and decompression in their seating. If the gasketry does not fit well and is not held tightly, gaps may open momentarily which let water reach the buried edges of the glazing or insulation panels or reach their seating. Then the oscillation of glass or panels and the pressure of the wind may pump or push water through to the inside. The more severe the exposure, the more robust the whole system, including glass stiffness, has to be to resist these mechanisms.

The following notes summarize experience in establishing causes of leakage, most of which results from poor assembly of factory-made insulation panels, bad detailing of junctions, and sloppy workmanship.

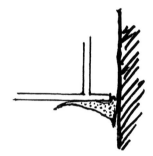

Figure 6.4a

- Poorly made infill units can fill with water from thermal pumping if their sealing fails.
- Mastic sealants are vulnerable and often leak but are nevertheless sometimes used as the only line of defence. This is unwise.
- They are also often used at joints not designed properly to receive them and this increases their vulnerability. For example, a bead of mastic may be left feather-edged on metal or glass surfaces (Figure 6.4a) where it will soon dry out and lose adhesion. Mastic should always have a proper recess to receive it (Figure 6.4b), and the junction it is protecting should not be subject to differential movement. The mastic should be tooled to ensure full wetting of adherent surfaces.

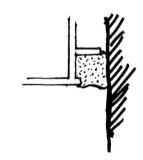

Figure 6.4b

- Where curtain walls abut brickwork or concrete, the materials are different and the construction tolerances greatly different. It has often been assumed that the gap can be adequately waterproofed by mastic, but this is improbable if the junction has not been designed for it. It is often the only line of defence, it is usually variable in width, and adhesion to brickwork or concrete is usually poor. Such junctions should always have a main line of defence by a DPC of some kind for which mastic only has to serve as an outer protection.
- Even factory assembly does not always enjoy good workmanship. A common defect occurs at the corners of framed windows or panel infill components where mitred joints have been found open after installation on-site.
- Occasionally one still finds glazing fully set in a bedding or glazing compound in the glazing rebate on curtain walls. This was an earlier practice but is not sufficiently reliable to meet the severity of exposed situations.
- Sometimes joints between components are designed to be sealed by a gasket of 'Christmas tree' section (Figure 6.5), pressed into place as the final stage of assembly. The act of pressing in anything of this kind extends it a trifle with each push, but gasket materials have a memory of what size and shape they were before the pushing and gradually pull themselves back to it, leaving the junctions open.

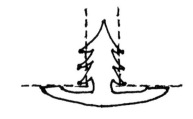

Figure 6.5
Christmas tree gasket.

If leakage occurs during curtain wall assembly, work should be stopped until the cause has been established, because it may not be possible to correct it later.

Case note

In one notable case leakage occurred soon after installation began. The cause was not established, remedial sealing by mastic was allowed, and assembly continued. This mastic also leaked and was covered by another mastic which leaked in its turn. This stage also leaked, and a third mastic was applied, in some instances over the previous applications, without success (Figure 6.6). Eventually the curtain wall had to come off and be thrown away, with much disturbance to tenants, heavy replacement costs, loss of rentals and so on.

6.2.11   Design and installation

Because the materials of which they are made and the technology necessary to accommodate thermal movement and exclude rain, curtain walls require specialist design skill and knowledge of manufacture and assembly. A choice has therefore to be made between proprietary and bespoke design, between full site assembly and delivery in sub-assemblies, between glazing from inside and from scaffolding and, of course, the character of facade desired.

6.3   Elastic systems

Most of the problems which have been described derive from the need to accommodate the contrary thermal movement of glass and metal components. The mechanics are complex and reliability is not always achieved. A logical approach is to cut out the mullions and transoms so that the only rigid elements are the glass components, and to hold and join these elastically. At present the most suitable elastic appears to be structural silicone, and Figures 6.7(a) and (b) show a proprietary system built up in this way. The glazed units have an elastic perimeter bonding, they are joined and sealed to one another elastically, and each panel is held at all four corners in elastic grips on metal brackets connected to the internal primary frame elements. The life of such systems will depend on the rate at which the elastics age, but the prospects seem good.

6.4   Testing

Proprietary curtain walling should have had comprehensive prototype testing before the product was marketed, and architects should see the certificate and test data to ensure that what has been tested is exactly what is being offered. Make sure that the tests were made with sample areas large enough to include all types of joints and components and that the weather conditions simulated were relevant for the climate area in which the building is to be built and the greatest severity of exposure to be expected. Dynamic wind tests may be necessary.

If in doubt about what is being offered, it is desirable to have an independent test done, and this will always be necessary in the case of bespoke design. Sometimes the assembly of the test specimen reveals shortcomings of detail design and the specifier should be present at the assembly on the test rig to look for these. Laboratories are available which will test specimens of adequate size under all necessary climate

Figure 6.6
Three coats of mastic applied in unsuccessful attempts to seal this junction.

Figure 6.7a

Figure 6.7b

simulations, 'adequate' meaning at least two storeys high and more than one module wide. At the time of writing there is no British Standard for such dynamic testing of this kind but the Architectural Aluminum Manufacturers Association of America has published a dynamic test regime (No. 501.1–83), and it is usual to specify compliance with it pending a European version.

Several factors can reduce the reliability of test results as a guide to actual performance *in situ*. Test specimens are usually better built than the real-life assemblies, and the transport and hoisting of unitized sub-assemblies and the site adjustments necessary when putting them into place may loosen or damage connections intended to be tight. Site testing should therefore also be done for water-tightness. At its simplest this is to spray water on an upper area and let it run down freely for 20 or 30 minutes or, preferably, use a hose to get some simulation of pressure effects.

6.5 Cleaning and inspection

Unlike stone and brick facades, curtain walls are not a form of cladding that weathers gracefully. To look well they have to be clean. It must also be possible to inspect them properly and cradle access is therefore usually necessary. A mild detergent can be used for cleaning.[2] The curtain wall contractor should provide a maintenance manual. Tactile tape protection should normally be provided for delivered metalwork.

6.6 Timber curtain walls

6.6.1 A timber curtain wall system

An elementary form of timber curtain walling came into use in the 1960s, mostly as infilling between brick or concrete cross-wall construction for flats. Its use has diminished but recurs from time to time. Some of the associated problems provided useful information about the use of built-up timber walling of any kind in an oceanic climate.

The curtain walling differs from the timber-framed house walls to be described in Chapter 7 in that it uses exposed stud framing as mullions and transoms with glazing and insulated infill panels set directly into it, somewhat in the manner of metal curtain walls. As originally developed, fixed glazing was usually held in place by small softwood trim planted inside and out (Figure 6.8), with sealing provided by a glazing compound. The trim was usually pinned in place, not screw fixed; it was often poorly primed or not primed at all, and it was frequently pushed tightly into place, squeezing the sealing compound to minimal thickness. One or all of the following then typically happened. The trim would absorb the volatiles from the glazing compound so that it dried out and lost its seal while the trim itself alternately swelled and shrank. Swelling on the outside was caused by rain run-off from large panes of glass, while inside it resulted from condensate, the causes of which were poorly understood at that time.

Metal windows were often used with timber-framed curtain walls. If the metal was steel its thermal movement seems to have been too small to cause problems, but the expansion of long aluminium frames could cause difficulties because it could force transom/mullion joints to open

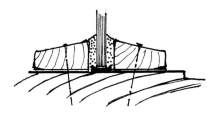

Figure 6.8

up, especially in prolonged hot, dry weather when the timber itself would be shrinking and the metal expanding.

A factor compounding these misfortunes was the use of reduced glass thicknesses. These were commonly the minimum thicknesses recommended in the relevant British Standard of the time in relation to the sizes of panes, but it seems that the thicknesses may have been set at the least dimensions necessary to avoid being blown in or sucked out by storm winds and not to avoid vibration. Whatever the reason, the glass was often thin enough to vibrate vigorously in wind – one could feel it easily by touch – and although it was usually structurally safe, the rapid bending that resulted around the edges of the glass helped to loosen the seal and to work the rain or condensate behind and under the trim where it then had access to the transom/mullion joints where there was the attendant risk of rot.

Planted stops are generally unreliable; it is wiser to shape the more substantial timber of the framing to receive the glass and then to seal it directly in place by a glazing compound. Experience suggests that with sensible detailing such joinery can expect a life of 30 to 40 years or more.

6.6.2   Boxing in

Infill panels, the 'opaques' for the timber-framed walls of the 1960s and 1970s, were usually boxed in by pinning relatively thin plywood, 5 or 6 mm, directly against planted spacer trim on the timber framing or by locking it in place against the spacer trim by beading, sometimes without sealants of any kind (Figure 6.9). The boxing suffered quite severely, sometimes becoming not only damp inside but also wet with some standing water. Cracking, delamination and general deterioration then took place.

There are the usual two ways by which dampness could develop inside such voids. If the inner face was plasterboard, indoor water vapour could migrate through it and condense on the inner face of the outer plywood when its temperature dropped below dew point. The other mechanism could be thermal pumping, external and indoor vapour being drawn in through leaky edges of the inner and outer panelling when the temperature in the void dropped, and then condensed on the back of the cold outer panel. Insulation would exacerbate the problems.

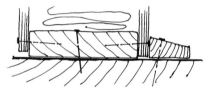

Figure 6.9

6.6.3   Maintenance

It is usual to specify external maintenance painting of all kinds at 3-year intervals, but there is much to be said in favour of doing the first maintenance on wood after no more than two seasons. By that time much of the initial restlessness of timber will have worked itself out, faulty sealing can be corrected, and other making good can be done. It is likely to contribute valuably to life expectancy.

A seldom-appreciated cause of paint film deterioration on timber is the use of sharp arrises. Although these make for a smart appearance, it is in the nature of the surface tension of liquid films that they thin out over sharp edges. This is where they are then weakest, and these become the positions where failure and peeling begin. The traditional specification of a 'pencil' round for arrises is apparently enough to offset most of this risk.

6.7   Some lessons

The experience of timber curtain walls has some lessons for designers:

- Be concerned about the quality of softwood to be used.
- Be sure that it has a quality preservative treatment.
- Don't mix hardwoods and softwoods or woods and metals without giving thought to their behavioural differences.
- Make sure that glass is thick enough for the pane sizes to avoid significant vibration in wind.
- Cut framing from solid wood rather than building up by planting trim whenever possible.
- Use microporous paint.

Reference

1   BS 476, Part 8: 1972. Fire Tests on Building Materials and Structures. Test Methods and Criteria for the Fire Resistance of Elements of Building Construction.
2   BS/CP 153: 1969 Windows and Rooflights; Pt. 1 Cleaning and Safety, and BS6262, Glazing for Buildings.

Further reading

1   Josey, B., 'Curtain walling specification', *EMap Architecture*, 1994. Valuable and comprehensive.
2   BRE Digest 73, Prevention of Decay in Window Joinery, August 1966.
3   BRE Digest 354, Painting Exterior Wood, September 1990. This is an excellent document.
4   Josey, B., 'GRP cladding specification', *EMap Architecture*, 1994. Glass reinforced polyester can be moulded and shaped into many forms, several of which find uses on curtain walls, chiefly as panelling. The article describes manufacture and technology relevant to its use.

7 Timber-framed walls for low-rise construction

I discussed in Chapter 4 stud-framed walls as an inner leaf of cavity walling where the outer leaf is masonry of some kind and the cavity contains insulation. This short chapter deals with self-finished stud walls, but because framing has been sufficiently discussed in the earlier reference only limited discussion of it will take place here.

7.1 Stud walling

Much of the detailing of modern stud-framed walls in Britain derives from North American practice where it has been the prevalent form of house construction for a long time. A rationale of modular timber sizing is pervasively influential there, 2-in. × 4-in. framing studs, 2-in. × 8-in. flooring joists, both of them spaced at intervals of 16 in. on centre to fit 48 in. wide sheet materials at three spacings per sheet. The inside face is usually plasterboard of the same size. The voids between the studs contain insulation, the outer face is sheathed in low-grade plywood, chipboard, or some equivalent for stability, and a decorative finish is applied, usually of rendering, clapboard, or clapboard substitutes. In Britain tile hanging would be added to such a list.

These elements have some relevant properties. Plasterboard is permeable to water vapour, although it can be reduced by foil backing if joints can be sealed, which doesn't happen often. Plywood can be assumed to be almost impermeable to water vapour because of the succession of gluelayers and is, of course, wind-tight. Renderings are also reasonably air- and vapour-tight, but weatherboarding in all forms and tile hanging offer little resistance to wind and none to water vapour.

There are therefore risks in following American practice in Britain without modification. Indoor climates in winter in North America have prevalently low or very low RH values, whereas in Britain they are relatively high. They will then also be high in the void from indoor humidity, so that if the plywood sheathing is impermeable, the insulation in the void will commonly cause its temperature to drop below dew point in cold weather and create condensate on its inner face. The sheathing is therefore a key element for consideration in the system, and there are probably two sensible options for its treatment in the UK's island climate.

The first is to make it permeable, for example by open jointing or perforation, with a micropermeable 'breather' membrane as a windbreak for the insulant. A rationale along those lines is arguably reasonable for air-permeable outer facings. Although the temperature of the sheathing and facings will still sometimes drop below dew point, only transient condensation would then be expected to form on them.

The second option is to use continuous plywood sheathing, but it must then stay above dew point temperature, and this would require a second insulant applied to its outer face, between it and the external finish. It should then be practicable to use either the wind-permeable outer finishes or renderings, but if wind-permeable finishes are used it would be desirable to cover this insulant rather than the other with a wind-cheater membrane of some kind to maintain its efficiency. A computation will probably be needed to check the best balance of insulation outside the sheathing and in the void between studs. It is customary for the main thermal insulation to be placed in the voids between studs.

7.2 Finishes

7.2.1 Rendering

If a rendered finish is used, the base must always be stainless steel lathing, as remarked previously; rust is otherwise going to cause trouble and premature failure in the UK's climate.

7.2.2 Plywood

Plywood has been popular as an outer finish in North America but it has not often been used in this way in the British Isles. Two housing estates have nevertheless been inspected where it was used and the results were informative.

Case notes

Although plywood is resistant to rapid dry/wet dimensional changes, on the first estate it suffered significant seasonal changes that caused trouble. Joints opened up and the subsequent wetting of the edges of the plywood initiated delamination. It may not have been marine ply, and its glue lams may not have been water and boil proof, (wbp).

On the second estate the plywood, 12 mm thick, had a softwood outer ply which had been spray-painted on laboratory advice with a colourless resin containing a decorative aggregate as a supposed protection. In fact rainwater and/or atmospheric vapour apparently diffused through the resin into the outer ply and there was possibly some gradual condensate dampening from behind. The grain of the softwood rose and delaminated and the paint peeled off. It all had to be covered over by another finish (Figure 7.1).

In the light of these cases and the general argument pursued in this chapter it is interesting that in a case examined by BRE[1] the moisture content of plywood applied as an outer cladding was found to reach levels of 27–30 per cent in midwinter, which is high.

Clapboard, whether of wood or a metal alternative, is bound to be air-permeable at joints, and tile hanging even more so. Decay of clapboarding or wall tiling battens seems never to happen. Tile hanging in particular has excellent longevity credentials both in cottage style and in the more sophisticated so-called mathematical tiling.

Figure 7.1

7.2.3 Paint finishes

The reader is referred to the discussions of paint finishes and mainte-nance for external woodwork in Sections 6.6.3 and 8.1.2.

7.3 Vapour checks

The recommendation which was frequently made in the 1950s and 1960s for a vapour check behind the inner plaster finish was always impracti-cal and is not worth attempting; it will always leak. In any case, as has been said here several times, vapour should be provided with a means of escape unless the indoor air is to be conditioned.

Case note

Some timber-framed houses were finished externally in dark anodized corrugated aluminium, and the flat roofs had an equally impermeable finish. Condensate formed in substantial amounts on the back of the aluminium and the chipboard undercladding became very wet.

7.4 Damp coursing

No problems need arise about DPCs for timber stud walls. The sole plates of the walls can normally sit directly on the DPC material as the cills of doors normally do, and the chief risk to avoid is leaks at the perforations of the DPCs for the fixing bolts holding the sole plates to the slab. These must be carefully sealed off. If the DPCs are bituminous, bitumen can usually do it.

If the wall sits on the edge of a slab on which a screed is to be laid, the DPC should have an inward selvage to which the DPM for the screed can be adhered. The DPC and the DPM should be of the same type of material to avoid incompatibility, as usual.

Damp sole-plates have been recorded in studies of timber construction in the UK. Two possible causes are drips of condensate from the plywood sheathing, and low temperatures due to heat loss through the slab edge.

7.5 Imported kits of parts

Countries where timber-framed housing is popular sometimes market kits of parts for complete houses as exports and some of these have found their way to Britain. Products of one such import have been examined.

Case note

The kits of parts were delivered in standard containers, one of which came on-site and was opened during my initial inspection. To everyone's surprise a lot of water ran out and, of course, everything inside was very damp. It was not clear why so much water had got in during the container's sea travels, but the upshot was that when heat was turned on to dry out the finished houses, warping was widespread. The windows in particular twisted to an extraordinary degree and broke up their triple glazing. Some 40 per cent of the windows had to be taken out and thrown away.

These houses were imported from North America, and a polythene vapour check was provided for the walls. It could not be installed without large gaps and could serve no useful purpose.

7.6 Conclusion

These notes should be read alongside the discussion of timber framed inner leaves of cavity walls in Section 4.8.3. There is no reason why timber stud walling cannot be used successfully in our oceanic climate provided that excess indoor vapour has a satisfactory means of escape.

Reference

1 Thorogood, R.P. and Rodwell, D.F.G., 'Moisture in the walls of timber-framed dwellings', BRE Current Paper 16/78.

8 Glass and glazing, doors and windows

8.1 Glass and glazing

8.1.1 Windows and style

For several thousand years the shapes of windows, and later of window panes, have been important stigmata of style, so when a fresh approach to architecture seemed necessary towards the end of the nineteenth century, traditions of window design were set aside as part of the process of breaking the historic mind-sets. Wright, Gropius, Le Corbusier and other heroes of the time consciously moved away from vertical openings to horizontal shapes and away from small sub-divisions to large panes.

The conventions of construction were much slower to change and as long as they persisted, the new thinking about windows and glass seems to have encountered few problems. This sedate situation did not survive the trauma of the Second World War, and when the pace of change began to quicken in the 1950s, different materials and techniques created new contexts which windows, glass and glazing often found uncongenial. Much was learnt from the problems that resulted.

8.1.2 Glass and framing

One of the earliest problems concerned the setting of glass in timber frames. This was discussed in Section 6.6.1 but a recap is convenient here. The usual practice, had been simply to putty the panes into place, but beading set in a softer, more adhesive glazing compound looked smarter and was presumed to give the large sheets of glass the greater freedom for thermal movement that they needed without loss of adhesion.

The beads were usually small and of softwood, given to much swelling and shrinking. They were often primed poorly or not at all, the glazing compound could easily be skimped, and steel fixing pins for the beading rusted and broke. The poorly primed beads sucked the volatiles out of the compound, leaving it hard and shrunken, and the movement of the glass then broke it up so that rain could get in behind the beading. Sometimes condensate seeped down inside, helping to wet the bottom member of the frame and getting into the corner joints where the end grain absorbed it.

Also, the quality of the joinery wood for windows and doors was often poor. Commercially grown timber rather than natural was being imported and marketed with no warning about its greater moisture movement and its corresponding inability to keep tight joints. Although it was more prone to rot it was sold without pre-sale preservative treatment. Paints were coming onto the market with an alkyd resin base to provide better coverage than before but they were also much less permeable, so that dampened wood was unable to dry out quickly as it had previously been able to do. Rot developed frequently, usually beginning at the joints, and replacement of the frames was sometimes necessary within only four or five years, a previously unheard-of rate of deterioration.

Occasionally when long horizontal panes of glass were used and set too tightly, their thermal lengthening in hot weather coincided with wood shrinkage so that corner joints got forced open. Sometimes, also, the glass used was too thin (See also Section 6.6.1). Recommended thicknesses based on pane size may have been sufficient to avoid blow-ins during storms but they were not adequate to avoid the vibration in wind to which large panes were subject. This not only helped to loosen settings but sometimes created a pumping action at the edges that enabled water to work its way around the glass to the inside, even erupting occasionally in little fountains along the bottom frame where most of it collected. Therefore, if you use beads,

• Use good wood, not too small, and screw fix it with non-rusting screws.
• Prime well both the beads and the frame to prevent the absorption of volatiles from the glazing compound.
• Keep about 3 mm of the compound either side of the glass to allow it to accommodate thermal movement without shearing.
• Use glass thick enough not to vibrate in normal storm wind.
• Give the glass space enough for thermal movement to take place harmlessly. Remember: the larger the pane, the greater the movement (See also Section 6.6.1).

In rehabilitating good Victorian or Edwardian housing one seldom has to throw away the windows: even after 80 or 100 years they are usually still in good condition. They are often of softwood and carry large panes of thickish glass. Some of Wren's hardwood windows at Hampton Court are still in good condition after more than three centuries. It is humbling to contemplate such longevity in the face of our post-1945 experience.

8.1.3 Thermal glass

Glass can be made heat-reflective by the infusion of gold or aluminium ions or it may be tinted to absorb solar heat and reduce the brightness of external views. In the latter case, the greater heat absorption will enhance thermal movement. When glass is to be used decoratively, e.g. as coloured spandrels, the colour coat may be fused on and it, too, is likely to be heat-absorbent. If a spandrel void contains thermal insulation the temperature range of the glass will be further enhanced.

Most of this thermal glass is likely to be set in metal or plastic frames – steel, aluminium or uPVC. Steel frames have moderate thermal movement and the others much more. If they have some freedom of movement it will at least be in the same direction as the glass and possibly of a similar order, but usually frames will be fixed firmly in brick, concrete, stone or curtain walls and there will be no 'give'. Consequently trouble can occur if the end of a pane abuts a projecting screw head so that its expansion jams it hard against this pressure point. The glass usually breaks. There must, therefore, always be enough room for its largest movement, and screws must never project into glazing rebates.

Large panes are usually seated on setting blocks, and if these get worked off to the side by glass movement the pane will ride unevenly and trouble can result. The blocks should be the full width of the glazing rebate.

Glass should not be used in ways that keep or allow the edges to get cold while the middle area gets hot. The hot area will expand but the cold perimeter will not and will form a ring in tension trying to hold in the expansion. The hot area will win and cause splits to begin along the cold edges which will propagate across the glass. Flaws caused in cutting are often the initiation points.

Case note

Double-glazing was improvized for insulation on about fifty skylights by setting Georgian wired glass on mastic tapes over the existing glazed skylight frames. As an elementary weather-protection for the tapes the glass oversailed them along the edges by 40 or 50 mm and the gap between the two glass layers was about 30 mm (Figure 8.1).

In early autumn the main areas of wired glass over the gap heated up quickly in morning sunshine while the edges stayed cool in chilly morning air. About half the wired glass fractured (Figure 8.2).

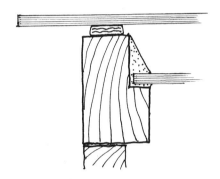

Figure 8.1

It might be thought that wired glass would offer enough resistance to cracking to avoid the problem, but not so. When it is being cut the glass gets highly stressed at the wire crossing points and sometimes incipient fractures occur which are not visible but open up when enough tensile stress is added to trigger propagation.

8.1.4   Toughened glass

There have been some spectacular failures of toughened glass over a long period. Toughening is done by heating pre-cut sheets slowly and then cooling them rapidly so that the surface shrinkage compresses a lot of stress into the body material. If the glass then gets broken, the fragments will characteristically have blunted edges due to the release of stress – a familiar example is when toughened glass windscreens break up. It can also happen on buildings and one cause which has come to light must be noted.

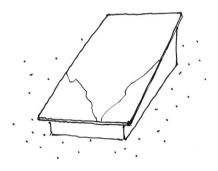

Figure 8.2

Case note

It seems that very small fragments of nickel, up to diameters of perhaps 0.3–0.5 mm, sometimes get into the melts of sand for the glass and these later slowly convert in warmth to become crystals of nickel sulphide. During conversion these expand and exert very high pressures, of the order of 30 000 psi or more, in the surrounding glass. This triggers the release of the in-built stress of the toughening process so that the glass breaks up. As a matter of incidental interest I have known the break-up to take up to 45 minutes, producing a grey noise that gradually diminishes. Sometimes an area around the crystal falls out locked in a coherent 'plate' or it may all stay in place until wind causes it to collapse (Figures 8.3 and 8.4).

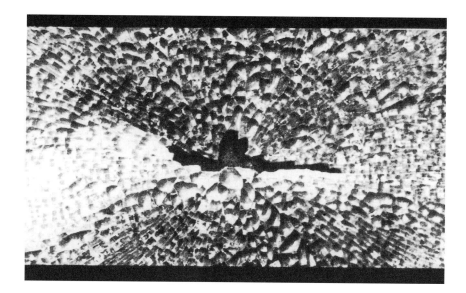

Figure 8.3
This is where the stone of nickel was.

Figure 8.4
A typical fall of a plate of the fractured glass. Note how the surrounding areas are distorted.

Most of the failures seen have been with tinted or colour-coated glass, presumably because it gets warmer than clear glass. There is no way that designers can anticipate or mitigate this problem. The industry tries very hard to prevent troublesome inclusions getting into the melts, but total success has so far been elusive.

This fault should not be mistaken for vandalism.

8.1.5 Alternative glazing materials

If glazing is done with clear plastic or a polycarbonate, it is essential to check its chemical compatibility with any glazing compound proposed for setting it or there may be damaging reactions. Also, note that its thermal movement is up to five times that of glass and that some grades of polycarbonate are harder and more scratch resistant than others. Be sure you get the material you need.

8.1.6 Double- and triple-glazing

Early makes of proprietary double-glazing gave a lot of trouble. That it worked at all was a triumph of technology over rationality, but in the end technology has surmounted most of the problems and the products have gradually become reasonably reliable. Their behavioural mechanics are as follows:

* The void is an insulant, so the inner pane stays close both to the indoor temperature and to its original size while the outer pane undergoes enhanced changes of temperature and dimension.
* The two panes have to be sealed together around the edges. At first this was done by bonding-in rigid spacer strips. A popular type was a square hollow tube perforated on the inner face and containing a desiccant to absorb any residual water vapour in the interspace air. Another was a solid lead strip with the interspace air evacuated and replaced by a drier air or an inert gas. Epoxides were often the bonding agent, but lead solder was also used.
* The difference in the thermal movement of the outer and inner panes naturally put a lot of shear stress into the bonding of the spacer strips and they often developed leaks.
* Air or whatever gas was in the void was warmed and cooled by the outdoor weather and therefore operated as a thermal pump. Thus when edge leaks developed, normal air containing a higher vapour content was sucked in cyclically and liquid water could also get drawn in if there was any in the glazing recess. Mist and condensate and scumming would then form in the interspace (Figure 8.5).

Gradually flexibility and sealing have been improved in the edge design and this has lowered the failure rate. In cold weather the outer glass may be drawn inward as the pressure of the interspace air drops (Figure 8.6).

Triple-glazing is now occasionally required and is usually done with additional hinged or sliding inner lights designed to seal well when closed, but openable to allow cleaning from inside (Section 6.2.7).

Glass is currently subject to much high-tech development, usually in the interest of smarter thermal performance, but speculation here will serve no useful purpose. So just try to be logical in evaluation.

8.2 Timber window and door framing

8.2.1 The frames of outward-opening units

A gradual shift from hand manufacture to mechanization of timber door and window frames took place in the inter-war and post-1945 periods as standardization slowly replaced individual detailing. Hand-manufacture and individual detail design had both been done in the context of traditions which prevailed because of their success in countering weather. However, memory of their value gradually faded as mechanization took command; the products themselves began to be used in

Figure 8.5
Failure of a lead seal for a double-glazed pane.

Figure 8.6
Distortion of the outer glass of double-glazing in cold weather.

more exposed positions on higher buildings in stylistically less protective architectural contexts. Also, in low-cost work, the wood was usually of the fast-grown commercial variety rather than the naturally grown and better behaved but increasingly scarce product. Eventually pre-treatment with preservatives became usual to avoid rot in the lower-quality wood, although it could not offset the greater moisture movement. The paint industry also restored some microporosity to its alkyd-based products to increase their permeability so that damp wood could dry out better.

The problems that began to arise as the traditions faded were mainly the following:

- Main frame joints began to give trouble.
- Leakage took place with increasing ease around opening lights.
- Leakage developed between frames and walls.
- Damage was caused by hardware.

Discussions of these follows.

Joints

Traditionally, frame joints were made by mortise and tenon, single or double. The tenon or tongue was usually put on the horizontal member and had parallel edges while the mortise was cut into the vertical member and had a slight outward splay. After checking for fit the joint would be glued and put together with glued wedges driven into the V-shaped gaps on the outer face (Figure 8.7). The tapering and wedges were early items to disappear as cheapness and mechanization took over, and loosening due to stress and shrinkage then became easier, allowing moisture to get in by capillarity.

Stub tenons were another form of economy. These provided less leverage and no opportunity for wedging. Sometimes nails were driven through the mortised member into the end grain of the stub tenon, probably to salve the conscience of operatives who knew that tenons should go right through. The nails could get no purchase in the end grain and tended to split the tenon or the mortised member, or both.

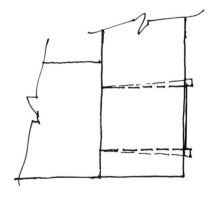

Figure 8.7

Case note

At one stage in this sad little history a rash of rot occurred on the top frames of outward-opening casements and doors due to the opening and closing of exposed joints where fast-growth timber had been used. The wood was untreated, the joint was made by stub tenon, and the impermeable alkyd resin paint was in use. The water soaked the end-grain wood and it could not dry. All the outward-opening doors of houses on one large estate lasted only a year or so.

In work of better quality, tenons were sometimes fixed by dowels, usually of hardwood, instead of or in addition to glue. In principle, dowelling is a strong technique but the old textbooks warn that dowels should only be used with well-seasoned framing wood because otherwise shrinkage and warping could snap them.

Why, then, did tenons set in mortises not also get snapped sometimes? An explanation might lie in the fact that the act of driving a snug tenon home would clean the glue off its outer end and concentrate it close to and on the meeting faces of the frame members. This could perhaps explain why, when shrinkage of the mortised member occurs, it does so towards the meeting faces, leaving the less well-glued ends of the tenons proud.

Comb joints were often used in place of the mortise and tenon because they are easy to make by machine and can offer a lot of leverage resistance. However, glue has often been meagre in the combs and then there is no strength; occasional examples have been seen without any glue whatever.

Case notes

On three package-deal design/build twenty-storey blocks of flats, imported semi-hardwood windows were used as sliders. The joints soon loosened by racking and leakage developed. Several windows were removed for examination and were found to have dry comb joints with a single star nail as a dowel. The nails had far too small a diameter to stand up to the racking thrusts of the opening and closing of sliders and were soon loose in enlarged holes. Damage to the combs and the glazing sealants made it too expensive to repair the windows and they were removed and destroyed. There were several thousand.

Another case of window failure due to weakened corners was differently instructive. It concerned some large and expensive imported windows, triple-glazed with in-built louvre blinds, and they were centrally pivoted horizontally for cleaning from the inside.

The frames of these opening lights had to accommodate in-built espagnolette bolting gear operated by a central lever on the bottom frame member. The frame had to be made up as box sections to receive the rodding, and their resulting size created an impression of robustness which was illusory, because the openings through the bottom corners for the rods and toggles left so little contact wood that a strong joint could not be made.

Unfortunately the windows were fitted by the builder and not by the supplier, for this shifted liability and lowered the skill level. Pivots for pivoting windows have to be aligned with high precision because misalignment causes twisting as the windows are opened and closed. These windows were misaligned and some warping also took place. The joints were too weak to cope and within a year or two many windows were inoperable, either stuck open or stuck closed. All had to be removed and destroyed.

8.2.2 Leakage between opening units and main frames

The gap between opening elements and the main frame is a specifically weak point for wind and rain resistance. Part of the problem has again been the poor quality of modern softwood which has necessitated a looser fit to accommodate swelling in wet weather; the looseness allows wind to get through more easily and take rain with it.

At the same time an apparently valuable tradition has become debased or has disappeared. This is the use of half-round grooves, cut in key positions on side-hung doors and casement windows and on the surrounding frame, and sometimes also on single- or double-hung sash. No-one seems to have been sufficiently concerned about their loss to research exactly how these contributed to success in the past and until such research is undertaken, we can only speculate.

On the side frames – the styles – a pair of half-round grooves about 10–12 mm wide would normally be cut opposite one another, and at the inner corner a single such groove would typically be cut in the outer frame. It is then reasonable to suppose that these would form expansion chambers so that a slit of pressurized wind carrying and pushing free rain into the gap between the two frames would expand into the chambers and lose energy, letting the moisture drain down to the cill. The groove at the inner corner would be a second line of defence working in the same way, and perhaps this was particularly necessary in areas with a high Driving Rain Index (Figure 8.8). Across the head the same detail was used for consistency in manufacture but it was seldom needed for weather protection.

At cill level things have to be different in order to drain off the arrested rain as well as the run-off from the glass. The cill was always sloped, but one or even two small steps were usually introduced on it as well, one of which was sometimes set back part way through to be in line with a vertical groove. In very exposed situations a weather-cast might be added externally across the bottom of the opening light to move the drip line farther out and provide a larger expansion chamber behind it to de-energize wind before it moved into the gap across the bottom. The stop-edge at the upper end of the slope clearly could also be more efficient in stopping rain when it had a groove rather than merely a straight upstand (Figure 8.9).

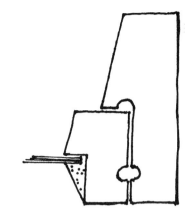

Figure 8.8

8.2.3   Inward-opening units

Windows and doors that open inwards have less natural protection than those that open outwards, and drainage from the weather grooving of timber frames is more difficult. It is therefore safest to use them in protected positions, e.g. under canopies or balconies, or set well back in reveals. This was overlooked in a number of blocks of post-war flats with balcony access. Front doors often had no canopy to shelter them and cill detailing was poor.

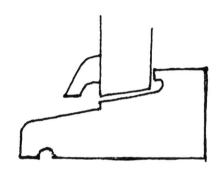

Figure 8.9

Case note

Some of these structures were of prefabricated concrete construction in which the floors were pre-cast beams. Rain got blown in over the door cill and found its way down through poorly sealed beam joints whence it dripped inhospitably inside the door of the flat beneath. The remedy was always expensive.

Inward-opening side-hung casement sash traditionally attracted a lot of ingenuity from architects working in this climate, sometimes with special metalwork gadgetry to supplement cunningly designed rebates. Given good timber and good workmanship (Figure 8.10) they were often successful.

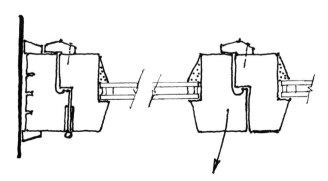

Figure 8.10
Inward-opening sash.

The cill and the style rebates have to be thought out together if the latter are to drain outward successfully. For example, if one considers using grooves on the sides of the frame to de-energize wind and rain, where should they be put so that rain running down inside them can reach the outward slope of the cill? This requires careful detailing.

Choosing windows and doors from catalogues is not always reliable; it is best to see some in use.

For doors there is a tradition of setting a metal water bar on the top of a timber cill, but it is simply another form of barrier in a rather inconvenient dirt- and damp-gathering position. The draught-stripping to which reference is made below should generally be a better solution. It has been somewhat easier to solve the inward-opening problems with extruded aluminium framing.

These are all lines of thought for designers; there are now no well-founded conventions to follow.

8.2.4   Catching condensate

After 1945, when traditional details had been simplified out of existence, there were numerous attempts to catch window condensate in some form of small gutter on the inside, with drains drilled outward through the cills. These holes could not be painted, they kept the wood wet and they clogged with dirt. Nothing of this sort was ever part of any tradition in this climate and was never needed. If condensate is occurring on this scale there is something else wrong which should be cured, not accommodated. (See Chapter 2.)

A perverse footnote: Why, when an outer door is to be in a sheltered position inaccessible to rain, is it still so often thought necessary to detail the cill for full weather protection when simpler and more elegant thresholds, less likely to trip people up, will do the job?

8.2.5   Draught-stripping

Numerous types of draught-stripping have been marketed for the DIY operator and they have the merit that if they arrest wind successfully they can also keep out rain. My own experience suggests that the best of them for this purpose is the spring strip of bronze or plastic (Figure 8.11). It has been around for a long time now, more than 70 years, which argues for its value. It is usually fixed to the rebate of the outer frame for the sides of doors or windows and to the tops and bottoms of the opening units. This avoids clogging by dirt at the bottom and makes tight corners possible at both top and bottom. It works well and without apparent deterioration. I have used it on new work as well as for remedials: it offsets the disadvantage of the looser fit timber framing needed for modern wood. It seems also to be more effective than other seals in preventing sound transmission via the air slots between opening lights or doors and their main frames.

It seems to me that spongey adhesive DIY draught-stripping intended to function by compression between opening light and sub-frame has a significant disadvantage. It can be compressed tightly where the closing handle operates but not elsewhere, and the opening frame then gets pushed outwards so that a permanent warp sometimes develops.

Figure 8.11

8.2.6   Single- and double-hung sash

Traditional single- and double-hung sash have a good record of keeping out wind and rain, but this depended on the use of good timber and good workmanship so that close fitting could be done while still allowing easy movement. With lower-quality wood a looser fit is necessary, but the spring bronze or plastic weather strip can be fitted and other devices are available. Any water getting into the side channels drains automatically to the sloping cill, and the sash can be dropped tight to the cill, so no weathering problem need arise there, although I think grooving should be done, especially in wet climate areas.

8.2.7   Horizontal sliders

Horizontal timber windows sliding on timber are not very satisfactory. Some form of metal or plastic slider is needed, and the bottom channel has to be drained. The jambs also need to be weathered and drained. The least satisfactory feature of sliders is that any ventilation needed will allow easy entry for wind-borne rain.

8.2.8   Pivots

Pivoted windows have the one notable benefit that they are easy to clean from the inside. However, because one part opens out and the other opens in, it is difficult both to weather the inward-opening part and to devise air- and water-tightness at the pivot itself. A test certificate should be sought for proprietary windows of this type.

8.2.9   Hardware

Hardware has an impossible task to perform when poor timber is used for doors and windows. If the fitting is done when the wood is swollen by a high moisture content in a rainy period, everything will be loose, rattling and leaking when seasonal drying occurs. If fitting is done when the wood is dry, damage by jamming and by forcing the hardware is more or less unavoidable when expansion comes in wet weather.

Case note

In one case the opening lights on seven twenty-storey blocks of flats were fitted and given their hardware in the factory where conditions were dry. When put in place, they soon could not be opened or closed without a hard push or pull. Almost all were soon irreparably damaged, usually by forcing the screw fixings of the hardware.

8.2.10   Frame-to-wall junctions

By far the largest use of timber windows and doors is in brick-and-block cavity walls. The greatest nuisance that appears to have occurred at the frame-to-wall junctions is water leakage through to the inner leaf and reveals. Sometimes it occurs near the upper corners of openings, or down the sides, or near or under the cill. Sometimes there is too much air leakage, enough to contribute to draughts.

This frame-to-wall junction is a risk position with most modern doors and windows, whether wood, metal or plastic. Treatment is discussed below.

Lintol

The conventional masonry practice until the 1970s was to use solid reinforced concrete lintols in which the depth needed to span openings was on the inner side, with a downward slope across the cavity to a thinner toe which carried the outer leaf. It had the merit of providing a solid bond between the two leaves and the disadvantage of being a cold bridge. Typically a crack also appeared on the internal plaster at the ends of the lintol due to shrinkage separation of the concrete from the inner leaf material, and the plaster surface on the lintol typically darkened by cold-bridge staining.

The slope across the cavity was intended to ensure discharge of cavity moisture through the outer leaf. It was customary to put a DPC felt on it and the toe to ensure discharge and so that any cracks in the lintol did not wet the inside faces.

Subsequently various forms of pressed steel lintols have come into use aiming to do a similar job, but offering a less solid bond between the two leaves while also forming less of a cold bridge. Other lintol arrangements have been used, and new types will come on the market because of the increases of cavity width resulting from increased insulation. What matters therefore is to identify in principle the design requirements that must be met at the heads of doors and windows:

- Insulants in the cavity over the opening have to be supported by the lintol.
- The inner and outer leaves need positive supported connection for stability.
- It must be possible to fix the head of the door or window frame to the lintol.
- Moisture getting into the cavity above the lintol must be discharged outwards and not into the cavity over the ends.
- The relation between any lintol DPC and the vertical DPC separating the inner and the outer leaf down the reveals must always give this junction adequate weather protection. A 3D sketch may be worth while for the builder.

The jambs

It is usual to return the inner leaf to the outer at the reveals of an opening and to separate the two by this vertical DPC. The returned stub wall should be associated with enough ties to ensure structural interaction between the two leaves down the sides of the opening. A thermal break may be desirable behind the DPC if the inner-leaf material does not offer much insulation, although the DPC itself, combined with shrinkage, will usually give a thermal break.

A dampness risk that recurs is the wetting of the inner reveal from one of two causes: water in the outer leaf getting past this vertical DPC by mortar/plaster contacts if separation is not complete, or wind-blown rain getting into the gap between frame and wall and reaching the stub wall. To counter these risks it is usual and sensible to require the vertical DPC to project sufficiently into the opening to prevent connection between the mortar of the outer leaf and plaster of the inner reveal. This DPC should at least contact the style of the frame and preferably engage a recess in it (Figure 8.12).

This can be the main protection against wind-borne rain, but the outer edge of the gap should be closed as an outer defence and wind-break. This is usually best done by a cover-strip. Mastic is often used instead of a cover strip on the assumption that the open edge of the gap must be made water-tight, but since the brickwork itself is not water-tight, its wetness will bypass the seal anyway, and absolute air-tightness may not be desirable because it can restrict the drying desirable for the back face of the timber frame. Simple closure of the gap seems to be what is needed, not a seal. Any sealing required should be done from inside, and a good plaster-to-frame connection is normally sufficient.

There was a long-established practice of making cuts 5–8 mm deep in the backs of frame timbers, sometimes said to be to prevent warping and sometimes to facilitate drying. Perhaps it is good for both and it seems a sensible practice to continue, even if the reason is not at present clear.

The outer reveals of classic Georgian windows were often rendered and painted white. The wall was solid, with the outer opening smaller than the inner so that shutters could be accommodated neatly on the inside. The windows were usually double-hung sash and their frame was offered up from inside against the outer brickwork. The arrangement gave good weather protection at the frame-to-wall junction.

Contemporary practice places door or window frames at various depths in openings, but whatever depth is chosen, the principles described are relevant in masonry walls. If it is intended that the DPC should engage a groove at the back of the frame, as is certainly good, the two must be positioned so that this can happen.

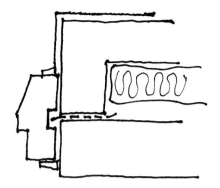

Figure 8.12

Window cills

The simplest form of cill is a widened frame member, wide enough to oversail the face of the outer leaf as a weather-cast. However, if the width is not to become unreasonable this implies that the frame as a whole is set in the outer leaf (Figure 8.13). Where this is done, two implications follow:

- The styles and head frame need to be set back at least 25 mm from the outer face to enable a cover mould for the frame-to-wall gap to be at least slightly sheltered rather than to be on the wall face.
- The whole of the frame needs to be wide enough to enable the vertical wall DPC to engage a groove in the styles of the frame.

The normal practice of builders is to bed cills directly on wet mortar; no harm seems to result if the cill is hardwood or has been rot-proofed. The brickwork beneath is unlikely to be very wet for prolonged periods because of the protection it gets from the cill.

When a narrower cill is preferred, no wider perhaps than the rest of the frame, one cottage tradition was to set up one or two tile courses as a weathercast on the outer leaf and to bed the frame so that the cill mastered the upper ends of the tiles and bridged the cavity to a raised inner leaf. This was a detail derived from the days of solid walls, but it works on cavities.

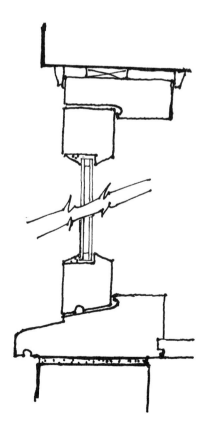

Figure 8.13

The chamfered brick cill which became popular in the 1970s may be a descendant of this detail or it may have been developed anew simply as a way of setting window frames deeply into walls. It needs to be done carefully to succeed.

In the most accident-prone version the chamfered bricks reached back to the inner leaf. Chamfers are used on edge, and, like all mortar perpends, those on chamfered brickwork do not get well-filled nor do they set tightly under compression as bedding mortar does. Therefore they easily take in water and the cills become very wet in rain for they also get the run-off from the glass above. For that reason, and because they were bedded against the inner leaf, it was necessary to place a DPC between the two and not only return it forward under the chamfers to the face of the outer leaf but also turn up the ends in the cavity to keep the water from running down the cavity face of the inner leaf or into cavity insulants (Figure 8.14).

But this was not the only risk. If the DPC was not supported across the cavity, the weight of the cill bricks and mortar squeezed it down so that it formed a shallow gutter which let the water run out at the ends into the insulation instead of discharging through the outer leaf (Figure 8.15).

A second, less troublesome, detail occurred at the top edge of the DPC. Again, it was not always recognized that the wetness of the chamfers could keep any mortar bed under a timber cill damp and that this could wet the inner cill and inner leaf. The upper edge of the DPC must project up through that mortar bed, preferably to engage a groove in the bottom of the cill to provide a positive stop.

Setting the window frame as far back as this implies that the cavity has to be closed at the reveals by turning the outer leaf back to the inner rather than by the normal and better arrangement of returning the inner to the outer. The vertical DPC can then line up with the back of the chamfer DPC. This deeper frame position also alters the relation of the vertical DPC to the lintol DPC at the top. Sill and lintol positions should both be checked by 3D sketches to make sure they are buildable and give proper protection.

A published detail a few years ago provides a basis for some final comments in principle (Figure 8.16). Comments:

- The opening light was simply rectangular without even a drip.
- The nose of the wood cill was thin, likely to shrink preferentially on the upper face and curl.
- The drip groove under the nose was shallow.
- The underside of the opening frame and the slope of the cill formed a funnel which would help wind to push water over the remainder of the cill.
- The DPC under the bedding mortar was unsupported and could not be held securely along its top edge. It would sag into a gutter shape when the bedding mortar was placed.
- It is difficult to see how mortar could be got into place behind the mastic bead or what function the rubber seal had.
- The mastic bead would also be difficult to place and would not seal well to the underside of the wood cill nor very well to the tiles.

One wonders what the builder made of it.

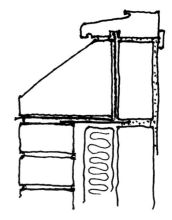

Figure 8.14

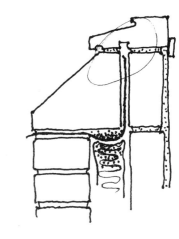

Figure 8.15

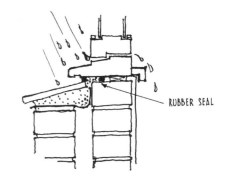

RUBBER SEAL

Figure 8.16

Door cills

Here we are concerned with what happens beneath cills. If the door is in a position protected from the weather, the only matter needing comment is that if it is at ground-floor level and seated on a slab or foundation that may get damp, the cill needs a DPC under it, normally part of a continuous wall DPC. At upper levels it should not be needed.

The problem is different where an unprotected door opens onto a balcony or roof deck. This is usually set down about 150 mm below the cill levels, so that its weather membrane can be turned up as a skirting to deal with rain splash. The height of this skirting is usually about 150 mm, and its top edge is either protected by a flashing or is turned inward into a brick joint or groove to be connected to a wall DPC.

The two most common deck roof membranes in Britain are still asphalts and roofing felts and these will serve to illustrate the points that need to be made.

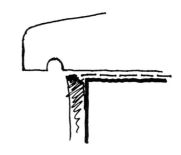

Figure 8.17
No room for a tuck-in.

Asphalt

In principle, the skirting and DPC have to be continuous beneath the cill but there is a practical problem. The door frame will usually have been put in place before the weather membrane of the deck is laid and it is not easy then for operatives to do an asphalt tuck-in reliably beneath it. Sometimes they find the cill seated hard down on the DPC with no room for a tuck-in (Figure 8.17), or if the cill has been put on a mortar bed (Figure 8.18), this may have been squeezed so far forward that there is no recess for the tuck-in. A useful way to deal with such risks could be to fix a Z-section in metal or GRP under the cill to guarantee sufficient clearance to make a tuck-in (Figure 8.19).

GRP would be the better material for the Z-section because bituminous materials bond well to it. This can be important, because the risk with asphalt skirtings is the tendency for the material to slump, and adhesion to the Z-section could help to hold it in place. There are discussions about related matters in Chapter 9 on roofs.

Figure 8.18
Mortar prevents a tuck-in.

Skirting felts

The felt skirting situation is different. The cill DPC will have been put in with the wall DPC when the wall came up to cill level, and then the door frame with the cill. Obviously it would be difficult to turn the upper edge of the felt back into a groove and bond it reliably to the DPC in this position. A better solution would seem to be to use something like a length of aluminium tubing to squeeze the felt back into the groove (Figure 8.20), but a Z section would probably be needed to control clearance under the sill.

Proprietary membranes that are not stuck down such as PVC, butyl and EPDM each have their own ways of forming skirtings and the specialist firms concerned should be consulted to ensure that they can adequately protect against splash-back getting behind them in the under-cill position. It can be difficult.

Figure 8.19
A Z-section to effect a junction.

8.3 Metal and plastic doors and windows

8.3.1 Steel, bronze, aluminium and plastics for framing

Steel, bronze, aluminium and plastics are typically used for a range of frames much smaller in section than those of timber, but the materials themselves impose constraints in the way they are used. Steel and

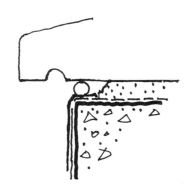

Figure 8.20
A possible tuck-in method for felts.

bronze sections are commonly hot rolled, although sheet steel is sometimes cold pressed. These processes result in simple shapes, and the strength needed comes from the nature of the metal itself and the relative thickness of the sections if they are rolled. Thickness also enables corners to be solidly welded even if the overall sizes of frame sections are fairly small. Thermal movement is low.

Aluminium sections are extruded. Their walls are relatively thin and the metal is weaker, so the strength needed is obtained by box shapes and larger overall dimensions in the section design. Complex sections are easily extruded. Frame corners are usually mitred and locked together by screw-fixed cleats or by crimping rather than by welding. If the extruded strip has a PVC coating this has to be welded at corners, usually by hot-air jet, to seal it across the mitre cuts. Otherwise it will shrink and open up. The final appearance of the PVC corners depends a lot on the skill used in this process and its finished appearance should be carefully checked.

Remember that thermal movement of aluminium is about twice that of steel.

The plastic used for wholly plastic doors and windows is uPVC. The 'u' stands for 'unplasticized' and reflects the fact that PVC in its familiar flexible sheet form was made so by plasticizer to the extent of 50 per cent or more of its content. Within the industry it was sometimes known irreverently as PVC reinforced plasticizer. It was a considerable chemical achievement to make unplasticized PVC extrudable and this perhaps prompted the retention of the distinctive 'u'.

The section shapes resemble those of aluminium for similar reasons, although aluminium or galvanized steel cores can be incorporated for added strength.

The thermal movement of PVC is greater even than that of aluminium and therefore considerably greater than that of steel.

8.3.2   Seals

At the time of writing the two most usual forms of seal are pile weather strips, known as brush seals, and solid, flexible tongue-like strips extruded in an enormous variety of special shapes, held in place in three or four different ways. The brush seal is siliconized woven propylene set on a variety of continuous bases shaped for insertion into reception grooves of various extruded or rolled shapes. The extrusions commonly use a hollow T-shape as a trap into which a matching solid T on the weather strip can be inserted. However, rolled steel can only achieve an open U-shaped recess, and a Christmas-tree section may then be used on the weather strip.

The solid flexible strip-seals are mostly made of butylene, neoprene or EPDM, although others are coming onto the market. They may be held in ways other than those just described and some of these look rather precarious although they seem to work fairly well. Two are illustrated in Figure 8.21.

The door and window frames can be expected to outlast the durability of all such aids to draught-proofing, so the question arises of their future availability for replacement. When they are no longer stocked it is not expensive to have a fresh die cut and new extrusions made, but one can imagine some inconvenience when the demand becomes intermittent and the jobs individually small. Will maintenance then be properly done?

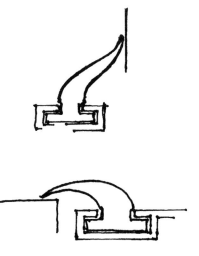

Figure 8.21

8.3.3 Thermal breaks

Most manufacturers of aluminium sections have also introduced split extrusions where one side of the frame is exposed indoors and the other outside, and they are held together by a moulded material which has low thermal transmission. The ostensible purpose is to insulate the indoor face in cold weather to avoid condensate forming on it. However, if condensate forms on the inner face, as I have said several times, the indoor water vapour levels are running too high (Chapter 2), and the proper approach is to get it under control.

It is naturally argued that the thermal break will also reduce energy loss, but the material used for it at present is not a very good insulator and the metalwork concerned is a very small proportion of the total area of the building envelope, so the saving is unlikely to be significant. It can be checked by a services consultant. Thermal breaks are more useful where there is colder winter weather.

8.3.4 Frame-to-wall junctions

While the windows and doors themselves should by now be able to be acceptably air- and water-tight for reasonable lengths of time, the trade literature is disturbingly deficient on good frame-to-wall construction. Consider the following four examples.

The first (Figure 8.22) shows a steel section set in a brick reveal. There is no DPC between the outer and inner leaves and rain has only to travel through a few millimetres of brickwork to wet the inner plaster, which it would certainly do.

The second (Figure 8.23) shows a frame set under a concrete lintol. The frame incorporates a thermal break but the concrete offers a continuous cold bridge. The mastic seal shown between frame and concrete could not be formed because there is no back-up for it. The mastic would just go through and not bond to the sides.

A third, under a timber lintol, showed a similar mastic bead but with the added disadvantage that wood, whether painted or not, is an unreliable surface on which to try to get durable adhesion.

A fourth was the only detail found in my mini-survey which recognized that a DPC must be present in a jamb, and that it must relate positively to the frame if internal surfaces are to be protected reliably from dampness coming from the wet outer leaf. However, there was a

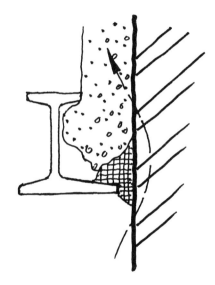

Figure 8.22

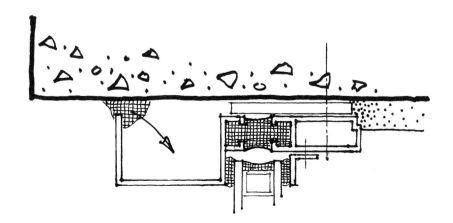

Figure 8.23

fillet of 'waterproof' mortar in this example which would serve no purpose. It would not bond to the metal and it would develop shrinkage cracks, both of which could become capillaries transmitting moisture. It would probably fall off soon.

The ignorance about good building practice displayed in these examples of proprietary product advertising is wholly unsatisfactory, although it helps to explain why so many kinds of building products get used in situations unfavourable to their performance. This often comes to light in litigation and indicates a less than healthy state of affairs in the construction industry. Building product makers should know more about good building construction than many of them seem to do.

Jambs

Frame-to-wall connections are as important for metal and plastic windows as for timber, and failure to think them through can cause trouble – wetting of the inner reveal, wetting of wall areas around the opening, and wetting of insulation. A vertical DPC in a jamb should be designed to be mated to the door or window frame in some positive way, as described for timber frames. It is an important protection.

Frame makers often seem to assume that frames should not fit tightly into openings and some specifically recommend a 3 mm gap, although no reason is given. If the gap is for thermal expansion this is not the way to do it because there is no slack in the fixings anyway. If it is to give a purchase for a bead of sealing mastic, it is not the way to do that either. Aluminium has the problem that the alkalinity of cements can damage it, so this has to be avoided. The most rational reason is that metal windows often go into place after masonry openings have been built, and because templates are not often used and masonry is not always very accurate, the gap provides a safety factor for clearance.

Domestic windows have traditionally been built in with the wall construction, and if this has been a cause of trouble, I have not heard of it. Probably the normal small imperfections of building construction do whatever is necessary.

For long runs of continuously connected windows some allowance for expansion must be made without leakage between the individual frames. Manufacturers usually have techniques for this.

As for mastic beads to seal frames to openings, consider first if they are needed. In masonry walls the DPC should be used to protect against dampness transfer to indoor construction, but one cannot generalize about other wall systems. If tidiness is a problem, consider whether a cover-strip will not be the neatest and safest treatment.

However, if in the end the choice is for mastic, a proper recess must be provided for it as described in Chapter 6, Figure 6.3, or it will not work anyway. The recess must be wide enough for proper injection (at least 6 or 7 mm), but need not be much more. There must be a back-up at a reasonable depth (no more than 8–10 mm) so that pressure of mastic on the sides of the recess will develop during application so as to get good adhesion. Ideally, the sides of the recess should be primed.

No frames of any kind presently available seem to provide for the kind of recess needed for mastic to do its job properly.

There is always the option of a rendered reveal, painted Georgian style. It is a good device in itself, it can look well in some modern contexts, and it is excellently protective with brickwork because rain is kept off the reveal at the point of greatest risk. It would not be desirable with aluminium.

If the outer cladding is brickwork and recessed joints are used, these should not be returned into window or door openings for this makes it impossible to form a reliably air- or water-tight frame-to-wall connection of any kind.

Window and door heads

The heads of openings to receive metal or plastic frames should offer no weathering difficulties, but there are two points to watch. First, the DPC which forms part of the lintol arrangement must discharge clear of the frame head. Sometimes it does not. Second, if there is a lintol with a smooth soffit outside the frame head, it must incorporate a reliable drip near the outer edge. Rain will otherwise run back on the soffit by wind and surface tension and may reach a vulnerable frame-to-soffit junction.

One instructive case of innovative window head design has been seen.

Case note

Bronze-anodized aluminium windows in long rows were to be given prominence by being set forward from the face of the brickwork by about 150 mm. Their projection was to be mastered by a bronze-anodized aluminium hood formed in lengths of about 1.5 m with an upstand at the brick face and an upper edge turned in at the brick joint above the toe of the lintol. There was a matching downstand at the bottom of the slope (Figure 8.24) which was bedded in mastic into a channel in the head frame of the windows.

The end-to-end connections of the hood pieces were made over a metal strip shaped with a recess, allowing a thickened bed of mastic under the end-to-end joint. The two ends of the hood rather amazingly had screw fixings to both sides of this strip, so slippage did not seem to be expected (Figure 8.25).

What soon happened was that the screw holes were ovalized by the insistent thermal movement of the hood sectors. Several millimetres of movement was eventually taking place overnight, and the mastic was so pulled about that it became wholly ineffective. When shrinkage occurred at night in rainy weather, water could run down the groove and get into the top channel of the window frames. From here it found its way through perforations into the vertical channels in the side frames and thence out on to the inner cill boards.

The movement of the hood sectors also tore the mastic bedding in the head frame of the window and where the top edge of the hood was bedded to the wall DPC. Both positions then leaked.

Cills

At cill level for windows, the trade literature often shows pressed aluminium or steel or extruded aluminium cills in weather-cast shapes, usually supported precariously by mastic tape or by a mortar fillet, sometimes of so-called 'waterproof' cement. The upper edge of the cill tucks in behind the bottom edge of the frame and is usually sealed there by a mastic fillet. The bottom edge, the weather-cast, is shaped to be trapped by holding-down cleats or lugs (Figure 8.26).

The highest-risk positions at cill level are usually at the ends of cills. This is where rain gets cornered intensively under wind pressure and

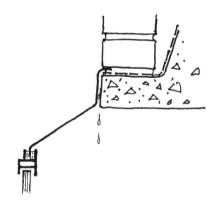

Figure 8.24

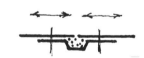

Figure 8.25

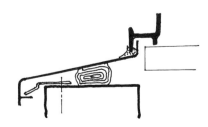

Figure 8.26
Precarious nonsense

where thermal movement of a cill has to be accommodated, with small gaps therefore opening and closing continuously. Steel cills will be less restless than aluminium, but regardless of the material it is a problem position for which there is no generally sound answer, although a turn-up end has frequently worked well enough. This needs careful thought.

Mid-span cill junctions remain a difficulty. All makers supply slip pieces, screw-fixed on one side only and usually intended to be sealed by mastic. This does not work. The mastic soon gets rolled inward or outward and the seal fails. It is probably best left dry with reliance placed on a long overlap rather than a sealant.

8.3.3 Points to watch about metal or plastic frames

- Sketch all corner details in 3D to make sure they work and can be built. Decisions based only on cross-sections often do not work at corners.
- Look carefully at the thermal movement of metal or plastic cills and consider whether it can be accommodated.
- As a generalization, provide a two-stage wind/rain protection between frame and jamb, usually combining protection at the weather face with a wall DPC well related to the frame. Neither is likely to be done very well *in situ*, so both should be used in moderate òr severe exposures.

No doubt architects will innovate well beyond what has been said here, but the principles necessary for successful wind and rain protection should be sufficiently clear to help towards success in innovation. Do not leave doubtful points for the builder to solve.

8.4 Testing for performance

As previously noted, test facilities now exist which enable the performance of windows, doors and whole wall systems to be thoroughly evaluated, and architects should seek test data on unfamiliar products and systems. If designers are innovating on their own account, they should make adequate use of these test facilities and allow plenty of time and an adequate budget for modifications.

8.5 Sound insulation

Weight, air-tightness and multiplication of membranes are all factors which can be exploited to increase sound insulation. A loose-fitting, single-glazed domestic window or door offers only about 15–22 dB sound reduction, whereas good air-tight double-glazed arrangements can get up into the 40s, and triple-glazing, done carefully, can reach 50 dB or so.

The width of an air gap is a factor in the amount of sound reduction obtainable because the transfer of sound energy across a gap takes place largely by elastic compression and decompression of the layer of air or gas it contains, and thin layers are less elastic than wide ones. Benefit does not improve much beyond 100 mm, however. A sound absorbent at the edge of the gap can reduce sound transmission significantly by absorbing energy in the compression waves. Good references are available.[2]

References

1 BS/CP153: Part 2, 1970. Windows and Rooflights, Durability and Maintenance. Replaced 1991 by BS 8213.
2 BS 8233: 1987; Code of Practice for Sound Insulation and Noise Reduction for Buildings.

Further reading

1 Josey, B., 'Curtain walling. Specification', *EMap Architecture*, 1994.
2 BS/CP153: Part 1, 1969. Windows and Rooflights, Cleaning and Safety. Replaced 1991 by BS 8213.
3 BS 6262: 1994; Glazing for Buildings.

9   Flat roofs

Contrary to some popular belief, the idea of the flat roof has been around a long time and has the most respectable of antecedents. The modern technology was not available to the earliest practitioners of Renaissance design in fifteenth-century Italy but they recognized from the outset the necessity of being able, if they so wished, to finish off the skyline of a classical building not by a pitched roof but by an entablature, with or without a balustrade, while keeping the roof behind it out of sight. Later in Britain when their time came, Inigo Jones, Christopher Wren and their successors throughout the seventeenth and eighteenth centuries all employed the idea.

The roofs themselves actually had shallow slopes but with the nineteenth-century development of the multi-storey cast iron-framed mill buildings the need was felt for true flat roofs and these became possible with the parallel growth of bituminous and asphaltic technology. In fact some mill owners astutely pushed it far enough along to carry a depth of 10 or 15 cm of water by which means they achieved several objectives simultaneously, storing water for mill use, keeping the top storey of the building cool in summer, and protecting the roof membrane itself from degradation by UV light, ozone and frost.

The roof decks of the mill buildings were usually thick timber but the advent of reinforced concrete framing around the turn of the twentieth century soon led to concrete decks on which asphalt and later bituminous felts were laid direct. This heralded the first serious misfortunes for flat roofs, blistering of the membrane and troublesome thermal movements of the deck itself, both of which were aggravated by the heat absorbed by the dark colour of the finish.

Between 1920 and 1940 the flat roof gradually became a symbol of the modern movement in architecture, but it is difficult for anyone who was not practising or not even a student at that time to understand how necessary it was to have such symbols. Debased Classical design prevailed, with a few notable exceptions, but it had been largely emptied of inspiration while the habit of thinking in Classical mannerisms for the limited range of buildings then thought to need architects inhibited the logical thought which was felt to be necessary if one was to make the best architectural use of new developments in construction technology for the widening range of building types required by modern society. The tiled pitched roof, successful as it had been in these islands, had to be able to be set aside for a period if designers' minds were to be cleared for balanced thinking about its own best use and about other potentially good forms of roofing. In the event, as was to

Figure 9.1

be expected, the pitched roof has again found a popular place but not the predominant position it formerly had, and not without its own new technical problems.

Unfortunately, reliable modern flat roof technology itself was not developed nor transferred to the design professions as rapidly as the situation warranted with the result that many failures occurred after 1945 and got such roofs a needless reputation for unreliability. They can be durable, efficient and attractive.

Most of the problems which have beset them since the 1950s have had a thermal origin related to the location of the additional insulation which began to be required. Sometimes this has been placed within the structure voids when the roof is timber framed; sometimes the deck top itself has been the insulant; sometimes insulation has been positioned above the deck, between it and the weather membrane, and sometimes it has been put above the membrane (Figure 9.1). Each location changes the behaviour of the whole roof as a system and in all positions but the last, the insulation becomes to some degree destructive. It is essential to understand the how and why of this.

In those parts of the British Isles in which snow in quantity is to be expected, remember that increased roof insulation allows greater loads of it to accumulate and sometimes they are very large (Chapter 1). There is no really logical way of dividing up a discussion of flat roofing because of the variety of ways in which the several types of structure interact with insulation in differing locations and with weather membranes of different types. The most convenient arrangement seems to be to put the focus in succession on the structural system, including insulation, then on the weather membranes, thirdly on boundary treatments, and finally on pavings.

9.1 The structure

The principal forms of support on which comment will be made here are the following:

* Timber framing
* Metal framing
* Concrete cast *in situ*
* Pre-cast concrete.

9.1.1 Timber-framed roofs

Timber-framed roofs typically comprise a joist system with a plaster or plastered plasterboard ceiling on the underside, a deck on the upper side carrying the weather membrane, and insulation in some form between the joists or forming the deck itself. The most usual materials for decking are plywood, blockboard or chipboard or, when insulated decking is wanted, woodwool, or subsequent competitors. Tongued and grooved boarding, once almost universal, has largely lost out, perhaps because of the high cross-grain moisture movement of modern softwood; if it is tightly laid it can expand and lift.

During construction, joists and decking may get either wet or dry before the building is closed in and they will then usually remain in their wet or dry state for some time, especially if the weather membrane is applied promptly. If they start wet and become dry the joist depths shrink and the deck joints open up, and the reverse if they start dry. Joist depths can alter by as much as 4–6 mm and panel dimensions by 2 to 4 mm, and warping is also likely and can twist the ceiling so that joints in plasterboard open. If the wood is poor, seasonal variations in humidity can keep this restlessness going.

When the building is occupied indoor water vapour will readily pass through the ceiling into the voids between joists and will typically condense in some position above the insulation, usually the deck/ membrane interface because it will often be below dew point due to the insulation. When timber roofs abut masonry walls or parapets it is usual to have a wood skirting as a kerb, but if this is fixed direct to the masonry while the framing and deck are free to make their moisture movements, any weather membrane adhered to both of them risks being damaged along the meeting line at the bottom of the skirting. In order to avoid this the skirting should be fixed to ride on the edge of the deck and not be attached to the masonry, and the top edge of the skirting must then be protected by a flashing (Figure 9.2).

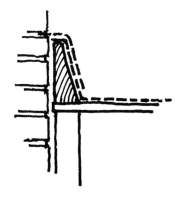

Figure 9.2

The opening and closing of deck panel joints can cause splitting and fatigue cracking in weather membranes stuck down across them and ways of minimizing or avoiding this risk have been developed and are considered later when membranes are discussed.

If strong panels such as plywood or blockboard are laid close-butted in a period of dry weather, seasonal increases of moisture content in them can build up substantial overall dimensional increases. A case is described in Chapter 10, page 268, where ventilated low-pitched roofs of chipboard, weathered in bitumen felt, built up summer-to-winter differences of 15–18 mm in an overall length of about 9 m and the same sort of thing can occur on flat roofs.

Deck panels of all kinds must be securely supported and fixed along all four edges and have enough intermediate support and fixing to avoid significant deflection when people walk about on the roof.

It is sensible to limit the greatest panel dimensions to about 1.2 m to avoid excessive movement at joints.

When insulation standards began to be raised in the 1950s the insulants were at first usually fibreglass or rockwool quilts placed loose in some way between the joists, often in association with foil-backed plasterboard for ceilings on the assumption that the foil would provide a vapour check as well as some further reduction of heat loss. Sometimes polythene was used as a vapour barrier instead of the foil but it could never be fixed to give a seal. In fact, none of these primitive ideas worked usefully. Water vapour seeped into the voids or was drawn in by thermal pumping and the insulation kept the temperatures of the deck and the air in the voids low enough for high RH values and condensate to form in them on chilly nights.

The next thought in what was a slow learning process was to vent the voids to the outside air in order to release the water vapour, but this is what happens in any case when the skirtings are not attached to a wall or parapet, and the nett effect is to allow the air and water vapour in the voids and the atmosphere to operate a continuous exchange in a search for pressure equalization, and circumstances will still recur which cause the deck and membrane to drop below dew point, causing condensate to form somewhere.

Many examples have been seen of the resulting problems and here are two of them. Their severity was unusual but they illustrate what one can be up against.

Case notes

In both cases the deck material was chipboard but not necessarily of modern roofing quality. Both took place on housing in which there was other evidence of higher-than-desirable indoor humidities.

In the first case the weather membrane was asphalt and condensate formed in the interface between it and the chipboard so that the top face of the chipboard expanded and caused the panels to arch between the support joists (Figure 9.3) while in the second case, where the weather membrane was aluminium-faced roofing felt on chipboard, the latter was wetted through and sagged badly (Figure 9.4).

Roof systems embodying insulation which causes the deck to stay cold like this in cold weather are commonly termed 'cold' roofs, though by the same mechanism they get exceptionally hot in hot weather. They have an obviously unresolvable dilemma and do not make good sense.

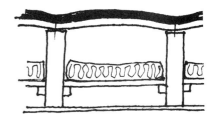

Figure 9.3

Figure 9.4
Sagging chipboard wetted by indoor condensation.

When the deck material itself is the insulant, the common material has been woodwool, and then different kinds of instructive problems have developed, either in the insulant or to the weather membrane. Woodwool is made of shredded softwood coated with a cement-based grout. It can survive a certain amount of cyclic changing of moisture content, but wood fibres embrittle and degrade when persistently dampened by alkaline moisture, and cement is alkaline. In a situation where RH values may be persistently high or where the woodwool is carrying a weathering under which condensate will form, it is at risk and sometimes degrades completely.

Case notes

In one case examined, 50 mm woodwool was set on a shallow slope 100–300 mm over a concrete structural roof on an air-conditioned building. Although it was a concrete slab, the case is mentioned in the present context because it was determined that it was itself too dry to be the source of the moisture later found in the woodwool, and it blocked off any indoor moisture, so the failure was unaffected by these risks.

However, the void was ventilated by standard roof vents, so humidity levels would be those of the atmosphere. The weather membrane was well-laid bitumen felt, and what happened was that the felt dropped below dew point on cold clear nights, causing condensate to form which dampened the woodwool. Over the relatively short period of a couple of years it degraded completely in several places and collapsed.

In another case, to which reference is made on page 262, the woodwool had a smooth cement finish on its upper face. It had been used to support lead which had to be replaced for reasons not related to the woodwool. The latter was carried on timber framing open to a much-used room below, and over a 20-year period it had embrittled to a point where operatives doing releading had to be warned of a breakage risk.

9.1.2   Metal-framed roofs

Metal-framed roofs are now usually decked in profiled metal sheet either carrying an insulant and a weather membrane or in some proprietary combination of them. They are discussed in detail later, but there is an important general point to be made here and it can best come from a case.

Case note

The roof was for a dry warehouse. The profiled metal spanned between steel framing Ts 1.2 m apart and was fixed to them by shot-fired nails in each alternate trough. The weathering system was built up by pouring some hot bitumen along the ridges of the metal – a lot of which ran uselessly into the troughs – with 25 mm extruded polystyrene roof board stuck down direct on to what remained if it had not by then cooled and lost its stickiness (Figure 9.5). The weather membrane was a two-layer built-up felt finish. There was no vapour barrier. A parapet nearly 1 m high surrounded the roof and the building was on a fairly exposed site near the top of a mile or so of shallow up-sloping terrain. It was said to have been designed to survive a wind of once-in-50-years severity. In fact within a year or two winds removed an area of the weather-proofing, insulation and trough decking near a corner. Repairs were made but in the following year adjacent areas of insulant and weathering farther from the corner came off but without the troughing. Failure recurred in several subsequent years, sometimes with and sometimes without the metal. The management instituted a policy of having maintenance staff walk the roof weekly to check for signs of trouble.

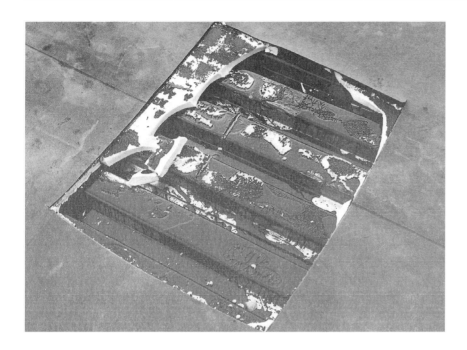

Figure 9.5

The problem was later analysed thus. The corner area suffered first because wind vortices concentrate with exceptional strength where the wind direction is such that it can come over adjacent parapets meeting at an angle. Wind is highly elastic (pages 6 and 7) and the vibrating vortices can pummel such a deck and cause it to flex vigorously, the corners of the parapet concentrating the energy from two directions. The flexing of the metal troughing in this case ovalized the nail holes so that the troughing lost an important contributor to its stiffness, and some of the nails actually fractured and disappeared. The bond of the polystyrene to the metal was erratic because so little bitumen had stayed on the ridges, or it had cooled before the insulant was laid, or because the flexing broke the bond or because the surface of the polystyrene was too weak and came away. Sometimes similar weakness at the top let the membrane alone come away. All these were probabilities with this unfortunate design.

Then, in the course of refixing the damaged areas, it seems that the repair men walking about caused so much deflection of the troughing that more of the fixing holes ovalized, allowing further wind-flexing to weaken or break the bond of the polystyrene in the adjacent area, and leaving the system vulnerable to subsequent winter storms. Damage was therefore able to spread like a toadstool ring. The weekly walking of the whole roof then made the damage general and the walks in fact became so noisy underfoot from broken polystyrene that the men came to believe that the roof had become too dangerous to be walked. When the deflection of the troughing was measured under the load of a man its mid-span sag was 30 or 35 mm.

By a coincidence one gale which damaged this roof destroyed another similar roof some 60 miles away. The insulant on this second roof was 50 mm isocyanurate panels but, like the polystyrene, they came off with the weather membrane and with some of the trough decking. The shot holes had again been ovalized and detached nail heads were widely scattered.

The key lesson here is that metal troughing used as a deck must be strong enough to span stiffly on its own between its supports so as to avoid deflecting more than a few millimetres under the load of a man carrying, for example, materials needed for maintenance, a total, perhaps, of 100–120 kg. Its deflection resistance should not depend upon its fixings, though these should certainly be in every trough to resist wind uplift.

Both of these roofs were actually designed by the makers of one of the products.

9.1.3 Concrete decks

Dense reinforced concrete decks cast *in situ* and screeded are typically 150–250 mm thick and have a very high thermal mass compared with steel or timber framing. They contain a lot of water initially and may require as much as six or seven years to dry to equilibrium moisture content, depending on their thickness and the drying conditions. Not much drying will take place outward on flat roofs because of the weather membrane but inward drying is accelerated if the building is power ventilated or air conditioned. During the early months the concrete will shrink and sag a little, as described earlier (page 134), and perhaps as much as 10–15 mm (Figure 9.6) on quite small spans, depending on the concrete quality and its reinforcement design.

It is usual to provide a screed laid to falls which slope sufficiently to ensure drainage in spite of the sag; slopes of 1:60 or 1:80 are the usual minima, but contrary to much popular belief, the chief purpose of drainage is simply to avoid messiness or to minimize leakage if a leak does develop. Ponding rarely harms bituminous or asphaltic membranes though it is said not to be desirable on plastic single plies exposed to freezing temperatures.

It is sensible to develop good adhesion between the screed and the base. A sharply rough surface left on the latter is a great help and if, just

Figure 9.6
Deflection of concrete roofs shown by ponding.

before screeding, the concrete is scrubbed with water containing a small amount of cement – not enough to be called a creamy grout – a good bond with the screed can develop. The water prevents the screed from losing too much moisture to the drier concrete, which otherwise would inhibit the interface setting of cement, and the little additional cement provides useful enrichment. It is a simple and excellent way of bonding new concrete to existing and was developed at BRE many years ago. It is not as well known as it should be.

Green concrete is really surprisingly flexible and if a screed can be adhered successfully to its supporting slab during the first few months of deflection, there is a reasonable chance that they will stay together subsequently.

Before there was much interest in roof insulation in this temperate climate, asphalt and built-up felt roofs were laid direct on the screed or the concrete base. The dark surface naturally increased the solar warming of the slab which then misbehaved in two characteristic ways. It expanded by simple enlargement and more complexly by swinging around anchorage points created by such stable parts of the structure as projecting lift or stair shafts or shaded areas, and if the slab was supported by steel or concrete beams, the heating of their upper edges made them arch. All these movements caused damage to walls and partitions of the top storey and sometimes twisted its structure (Figure 9.7) and it was this, rather than any perceived need to save energy, which in the early 1950s created pressures to find some way of incorporating insulation in the weather-proofing system to protect the concrete from solar warming. Appropriate insulants did not at first exist but when eventually they became available, *in-situ* concrete slabs proved to be an almost trouble-free type of deck.

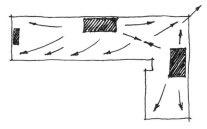

Figure 9.7

There is one modification of the cast-*in-situ* slab which is sometimes less than successful. This is formed with hollow clay blocks or other hollow formers in lines spaced so as to let concrete come down full-depth between them as ribs (Figure 9.8). The idea is, of course, to reduce weight while retaining the depth needed for stiffness. When clay blocks are used they have an undercut key on the bottom surface as a plastering base and narrow undercut tiles then provide continuity for plastering across the soffits of the ribs. The fault which can occur with this arrangement is that so much reinforcement may be needed near the bottoms of the ribs that not enough concrete gets through to cover and grip the steel or even to hold the tiles.

Figure 9.8

Case note

A case of this kind occurred in the ward block of a hospital. The soffits of the roof and floor slabs had not in fact been plastered because it had been decided instead to use a suspended ceiling to create a void for service piping. One of the rib tiles fell through the ceiling one day and landed on a patient's bed, and bare steel reinforcement could be seen. It was then found that other rib tiles came away easily and that the reinforcement then hung loose, making the floors unreliable and especially vulnerable in fire. In this case one of the pipe runs in the void was used for oxygen, by which a fire could be vigorously enriched.

This sort of situation – concrete unable to get through a congestion of reinforcing steel – has been seen in a reverse situation, and though the case did not concern a roof slab the same problem could arise on cantilevered roofs and deserves a mention.

Case note

Pre-cast reinforced concrete beams reached out from the building's internal framing to carry upper storeys by cantileverage. The cantilever beams had projecting stirrups along their top edges to bond to a floor slab which was to be cast *in situ* to provide additional depth which, together with tension steel going through the stirrups, was needed so that the beams could become adequate cantilevers (Figure 9.9).

Secretaries in rooms adjacent to the overhang eventually noticed that bits of concrete began to fall through gaps in louvred ceilings onto their typewriters. Some of the closed-in ceiling under the external cantilevered area was removed to have a look at the beams. What was then evident was that the slab concrete had not got through the congestion of tension steel so that neither it nor the stirrups were fully engaged in doing their job and the cantilevers were gradually failing.

Figure 9.9

This sort of thing is impossible to check during concreting and it may not even be visible when shuttering comes down, because of an outer skin of concrete that often forms. The one way of avoiding it is by understanding the risk, checking the steel spacing in the reinforcement drawings against intended aggregate size, checking the spacing on-site, and by requiring vibration of the concrete in relevant positions.

9.1.4   Pre-cast concrete roof slabs

Pre-cast dense concrete decks have the disadvantage that it is not possible to bond the pre-castings properly along their joints, and movement of the units and sometimes of the whole deck can occur. Reliance therefore should never be placed on decks of this kind to provide for windbracing. Drawings can show these joints well filled but in reality neither a good fill nor a good bond can be expected. Filling needs a very watery and therefore very shrinkable grout, and the sides of the castings will usually be too dry and dirty for reliable adhesion.

Case note

The roof of a large hall was formed of pre-cast panels set on steel beams. It was assumed that the jointed pre-castings would form a coherent slab such that no other wind bracing was needed. An unfortunate foundation movement at one corner of the hall caused the roof to distort and revealed that in fact it had no integrity as a slab. Instead the panels ground out the mortar between them as distortion took place, littering the floor below with chips of mortar. During rehabilitation the roof frame was properly wind-braced by steel framing.

Deck combinations

It can be unwise to mix deck types of contrasting thermal character for the same roof.

Case note

A case was examined in the United States where numerous plant rooms for a large laboratory building were provided with reinforced concrete platforms cast *in situ* while the remainder of the deck was metal troughing. Fibreglass roof board (25 mm) was laid over both types of roof and across the meeting line between them, and 20 mm of built-up bituminous roofing was laid over the fibreglass. Despite the insulation between the weathering and the decks, a thermal split developed along the junction lines (Figure 9.10).

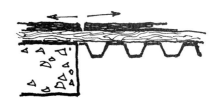

Figure 9.10

Plant room floors

Plant room floors are often built as part of the deck. Plant room equipment usually includes pipework containing liquids. Pumps break their seals, valves can develop leaks, and plant rooms can flood before the fault is noticed. In order to deal with such situations the floor and skirtings need to be waterproofed. Pipe penetrations also need to be treated so that flood liquid cannot slip through to rooms below. A drain should be provided, perhaps simply through to the roof, and door cills must not breach the waterproofing system, either of the roof or plant room floors.

9.1.6 Notes for designers

*   Always allow for vertical moisture movement of timber roof joists, as much as 6 mm.
*   Wood or chipboard panel decking must have up to 1 mm/m allowance for moisture movement. Laid damp it will shrink, laid dry it will expand. It will be seasonally restless.
*   Woodwool must be kept reasonably dry to avoid embrittlement.
*   Troughed metal decking must be stiff enough not to allow more than 3 or 4 mm sag under the weight of an operative, without dependence on its fixings.
*   Don't rely on pre-cast concrete decking to do lateral wind-bracing.
*   Where *in-situ* concrete decks rely on ribs for spanning support, make sure that concrete can and does get down well around the reinforcement.
*   Concrete decks sag significantly and permanently as they dry out.

9.2 Weather-proofing systems and materials

9.2.1 Locating the insulation

We have discussed the risks to be dealt with when insulation is placed in the body of the roof structure or when it forms the decking; let us look now at over-deck insulation. Potentially this may be placed under or over the weather membrane but the two positions produce dramatically different life expectancies for the membrane. Placed beneath it the insulation will shorten its life, sometimes to months, but if placed above it, the insulation becomes protective, and the membrane may have a long, trouble-free life.

The behavioural difference between the two has been dramatically demonstrated in a trial by Surman of the former Property Services Agency.[1] On a conventional reinforced concrete deck he placed two weather-proofing systems for a simple comparison. One was the initially popular sandwich arrangement, a 50-mm insulant placed on the deck and directly overlaid by 2-coat asphalt and the other, known as the protected membrane arrangement, with the asphalt on the concrete overlaid by the same insulant kept in place by 50 mm of gravel ballast (Figure 9.11). Surman ran his trial for six months to include exposure to summer/winter weather extremes and the following is a summary of his findings.

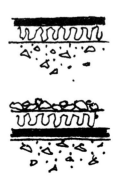

Figure 9.11

Insulation under the asphalt

The temperature in the asphalt overlying the insulant ranged seasonally from −15° to +43°C, a total of 58°C; the rate of temperature change in the asphalt was of the order of 30°C over periods of 6½–15 hours, and the temperature gradient through the asphalt was as much as 15°C at times.

Insulation over the asphalt

With the asphalt under the insulation its seasonal temperatures ranged only from +8.5°C to +24°C, i.e. 15.5°C, the rate of change was only a few degrees over several days, and the temperature gradient through the asphalt was less than 1 degree. So the seasonal range was reduced by nearly three-quarters and the asphalt never dropped near to freezing, clearly a much, much better regime. Changes in the gravel temperature had scarcely any effect on that of the asphalt.

Surman's findings are supported by a study at the US National Bureau of Standards.[2] A concrete deck was again the base. One of the test samples was an American-style built-up felted roof about 20 mm thick laid directly on the deck while the other was a similar membrane laid upon 25 mm thick fibreglass roof board. Thermal data were taken over a 10-hour day of summer sunshine and showers and the temperature curves are shown in Figure 9.12 for the ambient air, the membrane on the concrete and the membrane overlying the insulation. The latter provides the data relating to the present discussion for it will be seen that the membrane surface temperature rose to 130–142°F (55° or 60°C) three times in the day, fluctuating about 25°F (15°C) several times in a matter of minutes and dropping 60°F (35°C) in an hour and a quarter of rain. No data for a winter's day are available but in the Washington area −20°C on such a roof would not be unusual, making a typical winter-to-summer range of about 80°C.

In both experiments the high thermal capacity of the concrete deck was an unmeasured stabilizing factor but whatever its influence was, the

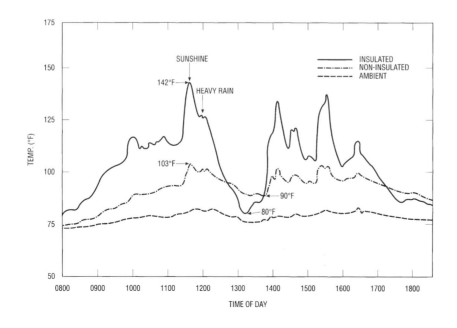

Figure 9.12
Time–temperature curves of insulated and non-insulated built-up roof specimens.

use of a lower thermal capacity deck could only increase the recorded variations.

With this background, then, I will look first at membranes overlying the insulation because it was mainly these which brought flat roofs into disrepute in the 1970s and 1980s and which in turn generated a welcome for the protected membrane arrangement. I will make occasional references to polymeric modifications to traditional membrane materials; these can perhaps best be taken at face value until the section on single plastic membranes is reached (page 211) where the nature of polymers becomes central to the discussion.

9.2.2 Insulation under the membrane – the sandwich system

Sandwich systems comprise a vapour barrier at the bottom, a weather membrane on top, with the insulant as the filling. We will take the vapour barrier first and then look at the membranes and insulants together because the characteristics of membranes determine some of the mechanical properties needed in the insulation.

Vapour barriers

In sandwich systems the vapour barrier* and the weather membrane must together wholly enclose the insulation in such a way as to prevent the entry of water vapour because this would lower its efficiency. The barrier must therefore have very low permeability the equivalent, at least, of a single bitumen felt of good quality, and be sealed reliably to the weather membrane around all edges, ducts, and pipes.

It is not known at present to what extent the efficiency of existing sandwich installations is diminished by vapour intake to the insulation but reports of energy loss significantly greater than predicted from a few evaluated buildings suggest that long-term deficiencies may be fairly widespread. Poor vapour barriers and seals have been so customary that this is to be expected. Thermal pumping could be the usual cause.

In the so-called cold roof situation described earlier (page 185), with insulation between joists in framed roofs, joints between deck panels of any kind are prone to thermo-moisture restlessness that can damage any fully adhered membrane, and protective part-bonding techniques had to be developed for asphalt and felted membranes. When an insulation sandwich overlies such a deck, the thermal and moisture conditions in the decking become more stable and correspondingly less likely to damage an adhered vapour barrier. If there is nevertheless thought to be a risk, the spot-bonding technique described later for felted roofs should be used for adhering the vapour barrier.

Where a deck is continuous a flood coat of bitumen can be used for bonding a bituminous barrier and will enhance its vapour resistance. If laps in the barrier membrane are to be made reliably, the base must provide continuous support. Illustrations in the trade literature often

---

*The terms 'vapour control layer' and 'vapour check' have come into use along with 'barrier' and they imply varying degrees of permeability. In view of what has been said about thermal pumping and the importance of keeping insulation dry, it does not make very good sense to drop below the quality of a true barrier when it is part of the enclosure for insulation.

show the barrier spanning unsupported over the troughs in profiled metal decking, for example, and this would be a recipe for failure. A well-secured flat sheet is needed as a base on such decking and in at least some trade literature this is recognized.

Edge seals for sandwich roofs

The necessity to seal off the insulation void not only along the roof boundaries but also around pipes, ducts or conduits that are found, sometimes unexpectedly, to have to pass through the insulation requires that the barrier and membrane materials must be chemically compatible – bitumens and plastics do not get on well together – and readily sealable. The barrier material in particular needs to be very workable because it has usually to be turned up and over to meet the weather membrane, and the work can be so awkward to do on-site and so accident-prone in workmanship that an effort should be made to get the number of pipes and ducts going through the insulant minimized and rationalized in advance as far as possible.

Circular pipes and ducts can be particularly awkward because there is no way by which a sheet material on a flat plane can make the transition to become a tubular surround geometrically intact. It becomes a scissors and paste job with a correspondingly reduced likelihood of success under site conditions. Pre-moulds should be available but do not always seem to be. Adhesive plastic sheet may seem sufficiently mouldable to do the job but plastic sheeting remembers the flat shape in which it was made and will try to return to it, usually with some success, and pull itself loose. Its memory is strong.

At roof boundaries it can be risky just to seal the two membranes to a wall or parapet because these are often not vapour-tight. Corners need to be made with pre-moulds because the folding or lapping of membranes is not reliable (page 94).

Weather membranes and insulants

Three classes of weather membrane will be considered, asphalts, built-up bituminous felts and single-ply synthetics, with a brief mention of flat metal roofs. The main duscussion on metal is in Chapter 10.

Asphalt

Asphalt as used for roofing is a viscous liquid, softening when warm and hardening, even embrittling, when cold. It is a mix of natural asphalt as found in various parts of the world – which is bitumen and particulate limestone, sometimes with a little clay – with factory additions in the form of acidic bitumen and finely ground limestone to extend its properties to those needed for roofing. In North America a similar product is called 'asphalt cement', but asphalt is not very frequently used in America in any form. In Britain and Ireland asphalt is normally laid in two, occasionally three, coats, each about 10 mm thick, over an open-weave bituminized fabric known as sheathing felt – a partial bonding material.

Sheathing felt was originally introduced to avoid the blistering which usually occurs when asphalt is laid directly on concrete. The explanation given for the blisters is that although adhesion might be generally good, there are sometimes small areas of non-adhesion which the vaporizing dampness in the concrete can exploit by expansion when the hot asphalt

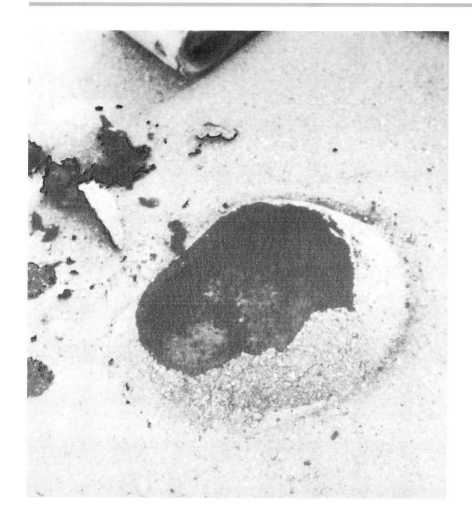

Figure 9.13
A fine asphalt blister.

is first applied and which then continues when solar warming of the asphalt takes place, lifting and stretching the warmed material into a shallow dome. But asphalt is not elastic; when it is deformed there is an internal rearrangement of the binder and aggregate which has no natural inducement to return to its previous flat state. When warming ceases therefore, the asphalt avoids a full return to flatness and then cyclic repetition gradually causes the blister to grow (Figure 9.13).

To deal with blistering the assumption was therefore made, back in the late 1920s, that a bituminized open-weave felt might prevent blistering by avoiding total adhesion anywhere, letting vapour disperse while yet retaining enough general adhesion to avoid uplift in severe wind. In practice it has worked well and blisters that still occur seem now to be due mostly to failures of adhesion between the two asphalt coats because of patches of dirt on the surface of the first coat, though it is also sometimes said that impurities in the bottom coat gasify when heated by sunshine and initiate blisters. I do not know any such cases.

Asphalt contains no reinforcement to provide tensile strength. It is held together simply by the stickiness of its bitumen binder and inter-locking in the aggregate. Sheathing felt probably was not originally envisaged as a reinforcement but it usefully performs this function for it has been found that if gaps occur between adjacent strips of it, the asphalt concentrates cyclic thermal movements along the gap line, gradually causing creasing and sometimes eventually splitting. It is always important therefore to form good laps in the sheathing felt, at least 75 mm, and to fix boundary edges firmly.

When asphalt is laid directly on a panellized cold roof, protection is needed against fatigue splits due to the opening and closing of joints. Sheathing felt can be a partial safeguard if the panel edge movements are restricted by good fixing, but the felt will often stick to the base and then movement is likely to cause damage. When asphalt is laid over an insulant there should be no likelihood of blistering caused by moisture but the sheathing felt remains necessary as reinforcement.

The composition of asphalt

A case from which very important lessons were learnt about asphalt occurred *in situ* on a reinforced concrete roof where the asphalt was laid in two coats on sheathing felt over 25 mm fibreglass roof-board.

Case note

In hot spring sunshine a few months after the asphalt was laid, widespread splitting developed alongside skirtings and along the stepped edges of shallow gutters and around upstand pipes, lightning arrester fixings and other fixed points (Figure 9.14). On hot days it also sometimes occurred in circular form around any place where a heavy person stood for a few minutes or where there was a concentration of foot traffic. So much water got under the asphalt that a walk across the roof could be surrounded by squirts of water 2 or 3 m away.

Eventually all the asphalt was replaced and to the same specification on the supposition that it had somehow been caused by poor workmanship. A different sub-contractor was selected for the renewal and the renewed roof suffered very little splitting.

However, the new sub-contractor used a different supplier's asphalt, and it looked quite different. Where the first had been almost silky smooth the replacement looked rougher and tougher. The difference in behaviour and texture prompted examination of the composition and this showed that while both asphalts were within the limits prescribed in the relevant British Standard,[3] the limits were wide and allowed one asphalt to have an unsatisfactory mix while the other was much more successful. The difference is very important. The British Standard composition limits at the time were as shown in Table 9.1:

Table 9.1

Material	% range in mix by weight	
	min.	max.
Soluble bitumen, used as the binder aggregate	11.0	13.5
Passing 200 mesh BS sieve	35.0	45.0
Passing 72 mesh and retained on 200	8.0	22.0
Passing 25 mesh and retained on 72	8.0	22.0
Passing 1/8-inch mesh and retained on 25	12.0	23.0
Retained on 1/8-inch sieve	0	1.0

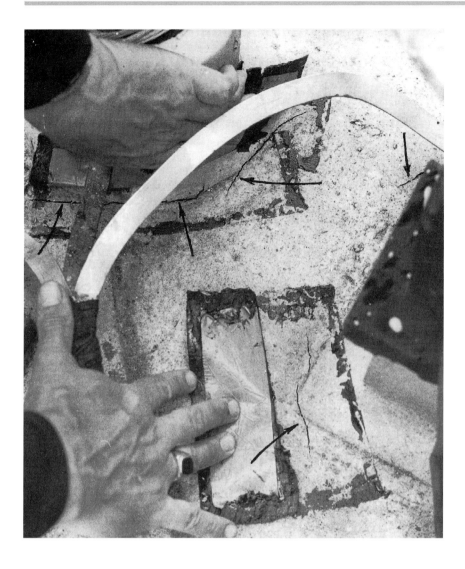

Figure 9.14

The asphalt which split so quickly and badly contained bitumen binder just on the low limit of 11 per cent, while the proportion of the smallest aggregate was at or over the maximum of 45 per cent and of the largest there was none, which was actually allowed. The intermediate aggregate sizes were in middling proportions. The predominance of fine aggregate and the absence of the largest size seem to explain the silky smoothness. The better and stronger asphalt, by contrast, had 12.5 per cent of bitumen, only 35 per cent of the smallest aggregate and 3 per cent of the largest (allowed at the relevant date), and middling proportions again of the intermediate sizes.

Now the facts of such a situation are, first, as explained earlier with concrete and mortar, that the greater the amount of fine material in any kind of bonded mix, the greater is the amount of surface area which has to be coated with the binder in order to do the bonding, and therefore the greater is the amount of binder needed. A deficiency of the latter when there is a high proportion of fine aggregate must therefore cause weakness because the coating of the particulate matter is then too meagre to bond it well. This explains one aspect of the failures.

Second, the lack of the coarsest aggregate deprives the mix of a valuable proportion of dimensionally stable solids. In principle, the argument again is, as it is for good concrete, to have a gradient of aggregate sizes from large to small in proportions such that they can come close to forming a continuum of hard solids needing only enough of the cementing agent to coat all their surfaces. The full range of aggregate sizes is probably therefore an important contributor to an optimal balance between adhesive and solids.

Other misfortunes have confirmed these general findings and it is unfortunate to have had a British Standard which could be met by two asphalts so different in performance over insulation.

During a later investigation another factor, mentioned briefly above, came to light. Natural asphalt does not of itself have the workability and other properties needed for roofing and it is extended by the addition of bitumen and crushed limestone to make up a finally desirable mix. It has apparently been known for a long time that the bitumen for this purpose needs to be acidic, presumably because the polymeric interaction of an acid with the alkali of the limestone improves the bond within the mix. I noted a similar interaction as the probable source of the unexpected durability of traditional sand:lime:dung renderings.

The bitumen is a manufactured material and the oil used in making it may be either naturally acidic or can be artificially acidified. Some within the asphalt industry claim that the naturally acidic oil, which apparently comes only from Venezuela, makes the better bitumen for the purpose and it may be significant that the asphalt which gave the good performance in the case just described came from a supplier using the naturally acidic bitumen as well as the better aggregate gradient.

The type of bitumen is not something which a specifier can control by specification but at least it should be possible when seeking asphalt tenders to require a statement of composition and this can be compared for probable performance with the composition of the two asphalts described. A sensible target is evidently a low percentage of the finest aggregate, a high percentage of the coarsest, and 12–13 per cent of the binder.

Throughout the remainder of this discussion I will refer to the latter type of mix as being the stronger or tougher asphalt and the other as being the weaker.

Reinforcement for asphalt

Sometimes the incorporation of expanded metal lath is recommended as reinforcement for particular positions but logically it could be argued that if this is needed, it could be because asphalt is being asked to do something for which it is not naturally suited and it may not then do it reliably. It has also to be borne in mind that asphalt reinforced in this way behaves differently from unreinforced material and some investigations have suggested that this has caused a weakening instability along the edge of the reinforcement. Bitumen felt has sometimes been applied to the edge areas of roofs for similar strengthening and in one case seen the asphalt developed splits intermittently along the edge of the felt. The fact is that asphalt likes physically stable and consistent conditions under, over or within it, or only gently graduated change.

When cracks occur in asphalt they often start so fine that they may not be visible though the asphalt has been fractured. Cyclic warming and cooling then gradually cause them to open and propagate. Many examples have been seen.

Insulants for flat roofs

The common roof insulants are foamed plastics such as isocyanurate, polyurethane and extruded polystyrene, with cork, resin-bonded fibre-glass, rockwool, and foamed glass as non-plastic types. Among the foamed plastics the isocyanurates are more stable thermally and may have better fire properties.

Asphalt on insulation

Asphalt laid in large areas on any insulation will develop maximal shrinkage stress in cold weather for which there may not be sufficient continuous restraint by the sheathing felt and insulant to allow the asphalt to remain intact. A split will then usually develop along one side or the other or occasionally somewhere in the middle, depending on the location of the best fixity.

Case note

On a university residence asphalt was laid on sheathing felt on a 50-mm layer of cork slab over a concrete deck. The cork was laid snugly jointed over a high-performance bitumen felt vapour barrier but was not stuck down to it, presumably because asphalt shrinkage was not anticipated. In fact in cold weather it did shrink but had sufficient adhesion to the insulant through its sheathing felt to pull the cork into compression by slippage on the barrier.

As it happened, the asphalt along one side of the roof was restrained by being under a line of paving slabs, so the shrinkage took the easier course of pulling it away from the other side, opening splits as wide as 20–25 mm along the base of the skirting in one cold January period. The roof was only about 8 m wide, so the shrinkage was about 3 mm per metre (Figure 9.15).

An interesting point was that iron soil vent pipes upstanding through the asphalt 60 or 70 cm inward from the skirting along which the splitting occurred offered enough anchorage to prevent splitting in its immediate vicinity (Figure 9.15). Fortunately the vapour barrier was so good in this case that no leakage occurred but the cork became so wet so that it was not much use as insulation.

Figure 9.15

It is essential that insulation be strong enough in compression to prevent asphalt being bent down by local loading such as an operative jumping off a plant housing, or the legs of heavy chairs or tables, or prolonged loads from planter boxes. In summer the warm asphalt can develop fine cracks which open up later if it is bent downwards under load, while in winter brittle fractures may be caused. Unfortunately there appear to be no UK strength criteria for insulants but there is a further discussion on page 205 in connection with built-up roofing.

Daywork joints

A day's work in laying asphalt has to end somewhere and along the finishing line it must be bonded later to another day's work. One of the main reasons – perhaps the only one that matters for laying asphalt in more than a single coat – is to provide for bonding by a lapped joint, usually of about 150 mm.

But a lap alone is not what matters; it is the temperature of the asphalt to which the bonding is to be done. Hot asphalt will not fuse properly with cold; it chills too much at the interface. Then, as it shrinks on cooling, the bond is easily broken. Cyclic warming and cooling will go to work on such laps and open them further, resulting in leakage.

And there is a second risk. It is usual to rub fine sand into the surface of the top coat while it is hot to prevent surface crazing. Sometimes some of it gets brushed or blown over onto the step formed for the lap and sets into the hot edge of the fresh asphalt (Figure 9.16) so that a good

Figure 9.16
Sand prevented a bond along this daywork-joint and leakage took place.

bond may be prevented when the next asphalt is lapped onto it. A bonding surface with sand on it has to be heated and the sand cut away to allow good fusing when fresh material is offered to it or there will be trouble.

The cheap and easy way of pre-heating asphalt for bonding is by torch but this can burn the surface without warming the body of the material and should never be used. The temperature of the existing asphalt should be raised by poulticing the edge with fresh asphalt until it is nearly as hot as the poultice, and then a good bond can be formed. It is a point to be specified and to be checked during site inspections because the integrity of the eventual membrane depends on it.

The same technique should be used for local repairs. The asphalt along the line to be cut should be warmed by poulticing and never, never cut cold by hammer and chisel; it simply causes cracking. The heated edge should be sliced to a slope for 100–150 mm so that a big enough bonding face is available for a strong lap to be formed, and if there is insulation beneath, it and the sheathing felt need to be made good before the repair asphalt is applied.

Hardness number

When asphalt has misbehaved it is usual to test it not only for composition but also for what is known as its hardness number. This should lie between 30 and 70 at 25°C. Somewhat oddly, the smaller the hardness number, the harder and more brittle the asphalt is. I have once seen a test result of Hardness 1 on a shiny, brittle, shrunken sample, indicating substantial overheating or the use of recooked material left over from an earlier day's work, or some other form of naughtiness.

Special and modified asphalts

There are special asphalts for such places as multi-storey car parks but they do not seem to present problems different in kind from those described for normal roofing, though they may differ in degree. The mix is usually composed to prevent softening under load in hot weather, more like a road asphalt.

Some road paving asphalts contain very large aggregate and probably rely to some extent on pounding by traffic to keep them in good order. They are unsuitable for roofs, though strangely enough I have seen one roof where paving asphalt, incorrectly marked on the blocks as roofing material to BS 988, was heated and laid on the roof without anyone being bothered about it. The material was extremely brittle and cracked badly over a considerable area when an accidental shock load fell on it and it all had to be replaced.

The industry has developed what are termed 'modified asphalts', to which plastic or rubber-based polymeric binders have been added. It is claimed that they give asphalt better performance and durability. The durability of asphalt is very high anyway unless it is compromised by misuse or in its composition, and it is not clear by what criteria performance is improved nor by how much. Elasticity seems likely to be increased.

This discussion of asphalt has been extended for a particular reason. Asphalt acquired a reputation in Britain for being the quality roofing in the years when there were no special insulation requirements and when it was usually laid direct on concrete, which provided it with a good heat sink to limit its temperature range and a stable base to restrain any restlessness. It suffered little and its natural durability came to the fore. Over insulation however, its behaviour changes so that although its potential durability is still there, its realization depends – and precariously at times – on the several kinds of factors which have been discussed here.

Built-up roofing

The term 'built-up roofing' refers in Britain to weather membranes typically comprising perhaps only one, usually two and occasionally three layers of bituminous felts, topped by a cap sheet or a crushed stone finish or some other protective surface treatment. The main felts used now are differentiated from less satisfactory predecessors by being termed 'high performance', by which is mainly meant that they have a stronger and more durable core – a polyester fleece or woven fibreglass – and that as with asphalt, polymeric modifications have been made to the bitumen to give the felts a more rubbery quality and to improve cold weather flexibility. Traditionally they have been bonded *in situ* by hot liquid bitumen but an alternative has been developed in the form of a contact coating which can be liquefied by a flame, and operatives lay this type by working with a long-arm torch, swinging the flame back and forth across the underside of the felt as they unroll it onto the roof.

When built-up roofing is to be laid on a jointed deck, a method is needed, as with asphalt, for partial bonding of the bottom felt to the substrate, sometimes to reduce the risk of blistering at the deck/felt interface and sometimes to avert damage to felts if they are bonded across deck joints that open and close. The sheathing felt used with asphalt is not appropriate with felts and instead a lower grade of bitumen felt perforated with quite large holes (Figure 9.17) is laid dry so that bonding to the deck, called spot-bonding, takes place only through the holes. Air and vapour pressures that otherwise might cause blistering can also then dissipate harmlessly, and the intermittency of the bond reduces the risks of damage along the panel joints. The size and spacing of the holes has presumably been designed to give sufficient bond to prevent wind uplift and to restrain shrinkage of the felt.

Spot bonding is the technique normally recommended when built-up membranes are to be laid directly on the decks of cold roofs, but is it needed over insulants in the sandwich situation? The risk of blistering is negligible but there is still a need to restrain both local and overall shrinkage on roofs. This means that not only must the insulants not be subject to shrinkage which can pull felt apart at joints but as with asphalt, there must be enough shear strength in the insulation and its fixing to resist the overall pull of shrinking felts, and unlike asphalt, felts are very strong in tension.

The strength of high-performance felts in tension can seem to be contradicted by the assertion that they can nevertheless be split in tension along panel joints that open up. The explanation appears to be that they are strong when tension is applied for a short period, as is done for standard testing, but time is also a factor and if tension is applied over a longish period, and especially if it is applied repetitively and is to some extent concentrated, failure can apparently take place at lower tensile loadings.

A cap sheet is typically a lower-weight felt incorporating a coloured mineral finish. This reduces access for UV radiation and restricts the departure of volatile ingredients, both of which cause degradation of

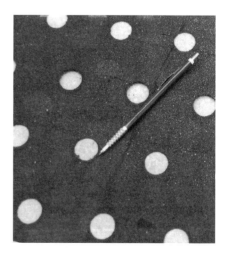

Figure 9.17

bitumen felts. Other surface dressings and finishes that give protection are discussed below.

Felts should always be stored at reasonable temperatures when they are about to be used because in cold weather even the most flexible types become stiff, and then incipient cracks can be formed by bending which will not be visible at the time but are likely to open and propagate later. A bituminous build-up relies a lot on being properly hot when it is put together, just as asphalt does. The trade recommends that when felts are being laid the hot bonding bitumen should not get more than about 1.5 m ahead of a felt to maintain adequate working temperatures, and the whole build-up should be made quickly to take advantage of the embodied warmth. Cold weather working must therefore be compensated by particular care, and if it gets very cold, work may have to be stopped.

Inter-felt blistering

Although spot bonding can largely avoid the cause of deck/felt blisters, they can still occur between felts and are especially common under the top felt – the cap sheet – because it takes the brunt of the weather. One cause, as with asphalt, is patches of dirt or dampness that prevent inter-felt adhesion but felting has another risk; the steps that are caused by laps may not be fully filled by the adhesive and instead air-filled runnels along the edge of the underfelt may get trapped and the air and vapour must then try to expand when warmed (Figure 9.18). If it does, it may cause peeling of the lap, which is easy compared with pulling it in shear.

Blistering is a pervasive nuisance on felted roofs, seldom causing leakage but looking as though it must. It is surprising that causation is not yet fully understood – reports of studies are very scarce – and that site practice has not yet reached reliability in avoiding the problem. As a matter of relevant interest one limited study at the National Bureau of Standards in the USA revealed the fact that the growth pressures for blisters can become astonishingly high, in one case reaching 400 psf.[4]

Figure 9.18

Pressures of this order are more than sufficient to cause the peeling apart of adhered membranes and stretching of a cap sheet.

The growth of blisters in felts differs in one or two obvious ways from what happens in asphalt. For example, felt lap slippage can take place when the felt is warm but the material is too weak to make it recover when cooling comes. Growth is quicker on felts, too, probably because the separated felt has a lower thermal capacity than a layer of asphalt, so it warms and allows enlargement readily but also cools and stiffens quickly in its enlarged shape and resists return.

Roofing is prone to unexpected happenings. A small forest of shrubbery developed on a roof comprising an overlay of felt on cork insulation (Figure 9.19). Presumably some seeds caught in the cork germinated in the warm humidity and, rather astonishingly, the growth was able to force its way through the membrane.

Figure 9.19

In practice, roofing is done at the mercy of the weather and usually under contractual time limits, both of which work against the quality of workmanship needed for this accident-prone part of the building envelope.

Asphalt and built-up roofing compared

The idea that asphalt is the quality roofing and that built-up felting is second-best dies hard, but we have seen that asphalt's reliability over insulation depends on its composition and that an asphalt may satisfy the currently relevant British Standard but may nevertheless be destroyed quickly, while bituminous felts can now offer quite reasonable durability and may outperform asphalt where tensile strength matters. It has therefore become more a question of which properties are wanted rather than of one material being inherently better than the other. In terms of breakdown by weathering, asphalt is the more durable.

Criteria for insulants

When felt or asphalt membranes are laid over insulation, the risk that this will magnify the thermal stress in these materials should now be clear, embrittling them in cold weather and softening them in hot, so insulants must offset these adversities by the kind of support they provide. The following are the principal requirements:

- The insulant must be well stuck down to the vapour barrier to resist sideways shrinkage of the applied membrane.
- It must not compress sideways and its surface must not give way in shear as the asphalt or built-up membrane tries to shrink.
- It must be fitted snugly so that there are no gaps at joints into which it can be pulled or pushed.
- It must be strong enough in vertical compression so that loads applied either in hot weather or cold cannot bend the membrane to the point of fracture. Typical loads are when maintenance staff jump down onto the roof from cleaning cradle gear or plant housings, or if they wheel loads across it.
- It should be thermally stable in its dimensions.
- And it must be undamaged by the heat of fresh asphalt or bitumen if these are applied.

None of these criteria appear to have had numbers attached to them up to the time of writing. They should be the subject of research with a view to an eventual standard.

Hot air spillage from air-conditioning plant

Cases have been seen where hot air or other gases have been discharged from air-processing equipment onto asphalt or felted membranes overlaid on insulation and it has damaged them badly.

Screeds as bases for bituminous membranes

Screeds have sometimes been laid over insulation as a base for asphalt or built-up roofing to overcome the compressibility of insulants and provide a firm base for the membrane. It was misguided and unfortunate. The screeds were unreinforced, shrank as they dried, and separated at daywork joints or other weak positions, and then moved about as large disparate slabs rather like the incipient spring break-up of an ice sheet.

Case note

The screed on several blocks of flats was laid over woodwool which in turn rested on polythene sheeting as a vapour check over pre-castings forming the deck. The screed broke up (Figures 9.20 and 9.21) so that water got in from above and water vapour also got in from below, passing readily through dry polythene laps. The woodwool disintegrated and the whole roof had to be replaced.

Surface finishes for asphalts and felts

In Surman's experiments (page 192) the surface temperature of the asphalt over insulation reached 43°C and the surface of the American built-up roof reached 60°C. Both figures could be higher on occasions,

Figure 9.20
The crack in a screed over insulation as betrayed by the rippled asphalt.

Figure 9.21
A broken and restless screed over an insulant, leading to splits in the asphalt.

and the thinner the membrane and the lower its thermal mass, the more extreme the peaks will be.

At such temperatures neither asphalt nor built-up feltings will be tolerant of much pedestrian use or concentrated loads like planters, leisure furniture, ladders or high heels. The materials indent and get messy, especially the weaker asphalts and lower-grade felts. Various top dressings and surface finishes have been marketed aimed either at keeping the membrane cooler – perhaps by 10°C or a little more – or spreading the loads. Some have worked, some have not, and some have actually caused damage.

Both white and silver paints have had their advocates to keep asphalt cool over insulation, but they have not gained much popularity. Their reflective efficiency does not last long and they are seldom renewed in routine maintenance. Other thermal factors can also defeat them.

Case note

Flat roofs of 100 mm thick metal-edged woodwool panels were overlaid by 20 mm asphalt with a silver paint finish. They were inspected when some were 8–9 years old and others about 10 years. Severe creasing had developed along panel lines in the younger roofs and splitting had begun on the older (Figure 9.22). The silver finish was not in bad condition but clearly the woodwool panels could still cause enough thermal ebb and flow of the individual areas of asphalt to destroy the roof along the panel joints in a decade or so. The composition of the asphalt was not checked in this instance.

Small crushed stone set in bitumen on built-up roofing has proved useful. It can look well but does not always adhere successfully. For good adhesion the bitumen must be hot, the stone clean and dry and

Figure 9.22

not cold enough when laid to chill the bitumen quickly. If the stone is not tightly held, water can get between it and the bitumen and freeze in cold weather, causing general loosening. Some of this will happen anyway but it should not be encouraged. If it gets too loose it can be blown about.

Why stone dressings provide protection is not clear and again there seems to be no reported research. One can speculate that it is a combination of factors, the exclusion of UV radiation, the avoidance of high temperatures by contained shadow and by the increase of cooling surface, the increase of thermal mass and the restriction of volatile losses. One distinctive gain seems to be a reduced incidence of blistering, though again the reason is not clear.

Of the types of stone used, white marble had a period of popularity in the expectation that it would stay looking white and keep the roofing very cool, but its adhesion seems to have given more than usual trouble – it's a dusty material – and some roofs soon looked very patchy.

Calcined flint seemed to fare better, adhering and looking quite well, basically white but with attractive touches of red. When I have used it, a thickish coat was laid with loose material overlying the adhered layer and sprayed with a clear adhesive around drainage outlets and for a width of a metre or so along the roof boundary to prevent wind scour. It worked well. But grey granite has become the norm, again looking well and seemingly successful.

Hard finishes are another matter. Two early types on the market were tiles 300 mm square, of white, rather porous decorative concrete about 25 mm thick, or compressed asbestos tiles, again white, but thinner. Both were intended to be bedded in bitumen and the custom was to set them about 3 mm apart. Unfortunately both had thermo-moisture movement of about 1 mm from hot/wet to cold/dry and when this much movement is imposed on a 3 mm joint width of asphalt or built-up roofing in cold weather when it has reduced elasticity, the internal adhesion breaks down and shrinkage of the tiles eventually pulls it apart along the gap lines.

In hot weather the behaviour changes, especially on asphalt. Any adhered tiling makes any large area behave more like a series of flat plates, pushing the asphalt membrane as a whole against any fixed positions when it expands and causing creasing such as that in Figures 9.23 and 9.24. It looks bad but is more often just symptomatic of what is happening rather than causing leakage. Recovery when the temperature drops is not coherent; it is every tile for itself, and splitting along some of the gaps results as described above.

Ordinary paving slabs, 50 mm thick and 600 × 600 mm up to 600 × 900 mm in size, are often laid close-butted and dry on asphalt which has insulation beneath it. They need no adhesive, but what is easily overlooked is that dirt and grit gets into the joints and the repetitive warming and cooling of the slabs lets more of it in so that lateral expansion of whole areas occurs, gradually and collectively, and often with unexpected force. Such paving has been found biting deeply into asphalt skirtings. Even if the pavers were laid with 5 or 6 mm gaps to start with, expansion damage has been found to take place.

Sometimes a border of loose gravel is put between slabs and skirtings to allow for the expansion but cases have been seen where crushed stone was used and this locked up and jammed in the border and bit badly into the skirting materials.

Figure 9.23

Figure 9.24

Troubles of these kinds encouraged the development of small stools to carry the slabs above the membrane. These have cruciform ridging that locks the slabs into place at corners as well as setting their spacing. But the weight of the slabs is substantial and in hot weather the small stools sometimes dig into the membrane and then get pushed about at the corners by slab expansion, causing some damage. In principle, the idea is good and probably what is needed to make it more reliable is to use thinner and lighter slabs, and if the membrane is asphalt, it should always be the stronger sort. The thermal mass of the slabs and the shade they provide both help to keep the membrane cool in hot weather, and dirt drops through the joints instead of consolidating in them.

A useful but underexploited surface treatment is duck-boarding. Made with good-quality timber it is reasonably durable and has no vices. It shades and protects the membrane so that what it may lack in its own longevity is compensated by the long life it gives to the asphalt or felt beneath.

Roofing felts are available with facings of thin embossed copper or aluminium, intended to offer a cap sheet finish with the image-appeal of these metals. Sometimes they have been laid felt-style with simple laps but more usually in metal roofing-style with upstand welts or rolls, though unlike 'real' metal roofing, where the welts embody holding-clips fixed direct to the deck (page 246), metal-faced cap sheets rely on the normal adhesion used for built-up felted roofs.

Bonding together materials which have very different properties always has uncertainties and this is true of metal-faced felts, perhaps partly because the thermal coefficients and strengths of felts and metals differ so greatly. The felt used probably has a coefficient of expansion about four times greater than that of the metal but often the latter proves strong enough to take charge.

Case note

On one flat roof roofed with a copper-faced felt there was evidence along the lap edges of thermal shrinkage by the metal which pulled the felt with it. Loosening of the bitumen adhesion of felt over metal also occurred and in the cold weather of the day of inspection it was also noticed that if one stepped close to a lap, there was enough compliance underfoot to allow the lap to open up by leverage. In some places the lap opened up and arched due to lengthwise thermal expansion of the metal and the gaps became sufficiently large to suffer further opening by strong wind (Figure 9.25).

Figure 9.25

The circumstances of manufacture may also have something to do with the uncertainty of metal/felt combinations because they are rolled for distribution, perhaps fresh from the bonding process, and acquire some set curvature, but have then to be flattened for use which would be expected to put some compression into the felt and tension into the metal. The appearance of some of the metal seen on sites seemed to confirm stress of this kind.

Among welted roofs of metal-clad felts seen at various times, one was particularly instructive. A central area of the roof was laid direct on chipboard to a nominal slope of 1:40 and around it the roof sloped steeply downward Mansard-style. The welts on the central area were folded flat to avoid walking damage by maintenance staff. There were several points of interest.

Case notes

First, drawing the metal-faced felt over to form welts apparently lifted and disbonded the adjacent felt on the side away from the fold, just as it does on conventional metal roofs (page 246). Bending the welts flat on the central area seemed to increase this.

Second, where flattened welts encountered transverse end-to-end joints, this was done metal-working style by beating, but on this thin metal it caused crevice cracking and some corrosion was found to have begun on the inside of some folds. (Crevice corrosion is discussed in Chapter 10.)

Third, the welts loosened; thermal movement seemed the only explanation.

Fourth, shaping and folding around the tops of skylights set in the Mansard slope seemed to have been too difficult to do reliably and leakage resulted.

On the flat part of the roof some water worked through the welts to the adjacent disbonded areas, perhaps drawn in partly by thermal pumping. Underfelting in these areas was found to have developed some fractures and leakage had occurred. Both the flat welts and the standing welts on the steeper slopes were loose and wet, probably sometimes from dew or condensate, and the wetness could also run down the slope inside the welt.

This case was not untypical of some other roofs seen which used aluminium or copper-clad felt and although no doubt successful roofs of this type have been done, it does raise the question as to whether the combination is disadvantaged by being amenable neither to true felt-laying technique nor that of metal working.

Single-ply membranes

A significant distinction has to be made between the discussion of bituminous roofings and single-ply synthetics. With asphalt and bitumen felts the decisions about detailing and specification are mainly in the hands of architects and the building trade while single-ply membranes – the plastics and synthetic rubbers – have now usually become parts of proprietary systems and are best used as such because of the mechanical and chemical compatibilities involved.

However, architects and the building trades are inundated with terms like monomers, polymers, elastomers and so on without being given much idea of what they mean or imply for the behaviour and durability of these products. For that reason my first discussion of plastic plies is devoted to building up a small information bank as a base for choice and for architects' advice to clients.

Single-ply synthetics laid over insulation are laid dry and usually held down by mechanical fasteners fixed through the insulant and vapour barrier to the deck, or by ballasting, or by adhesive. The membranes are products of polymer chemistry. Logically enough, polymers are formed from monomers, which are relatively small individual molecules, usually of a gas or liquid, which are linkable to form what are called polymer chains. A useful analogue is linking paper clips end-to-end, and just as paper clips of various types can be linked and cross-linked in a variety of ways, so can monomer molecules and some of the results may then get called co-polymers. Molecular manipulation of this kind together with the use of plasticizers and additives is actively exploited to produce polymeric products with a variety of engineered properties. Elasticity, flexibility and fatigue resistance are examples of what would be sought for roofing.

Some polymeric products are naturally rigid and are made flexible by a plasticizing additive. PVC – polyvinyl chloride – is one such. But plasticizers apparently do not combine permanently with the other ingredients within a product. Some are also volatile to a significant extent and over the timescales relevant for buildings they can migrate or be absorbed into other materials with which the product is in contact, or they may simply evaporate or be washed away gradually by rain from exposed surfaces. In a building context migration may take place into mortar or into chemically contrived products such as extruded polystyrene, and sometimes manufacturers warn about this risk. They refer to it as 'incompatibility' and what they usually mean is that their particular product may be destabilized by it. Instances have been known of timescales as short as a few months for this.

The loss of any ingredient in a product depletes its volume and causes it to shrink, and the loss of a plasticizer will specifically also return the product slowly towards its rigid state which, in the case of a thin membrane such as roof sheeting, will mean embrittlement and fragility.

Some plasticizers in materials under pressure, such as DPCs, are known also to flow gradually away from pressure points to nearby unpressured areas. The depletion in one position again can cause shrinkage due to loss of volume while the enriched areas nearby may be destabilized and weakened by becoming overplasticized. The two mechanisms seem typically to produce different forms of cracks or splits and the appearance may change, the depleted material losing gloss and the enriched areas getting glossier.

The stability of non-plasticized material is usually much better. Survival depends on the processes of oxidation and the propagation of cracks being sufficiently slow. Suitable stabilizers and additives have been developed to delay these processes.

The terms 'thermoplastic' and 'thermoset' will be encountered in references to plastics. It happens that some polymers – those classed as thermosets – are set permanently by heat, often during formation; epoxies are examples. The others, known as thermoplastics, resoften at fairly high temperatures and then return to their previous state on cooling. The latter can therefore be welded by heat at laps, often done by a hot-air jet, while thermosets have to be bonded by adhesive or a solvent.

The terms 'rubber-like' and 'elastomer' tend to be used interchangeably. The products are naturally flexible and to some degree elastic, and development of the chain structure of thermoplastic polymers can sometimes give them rubbery qualities so that in practice product differences between some plastics and rubbers are diminishing.

The term 'elastomers' sometimes turns up to describe polymerics with elastic properties and these are apparently complex cross-linked combinations of large organic molecules. They are synthetic rubbers and, like rubbery products in general, they will deform under stress and subsequently try to recover. They tend to lose some flexibility at low temperatures and creep may occur when they are hot.

The various polymeric combinations are given names that attempt to balance market memorability with chemical identity. Some examples are the PVC just mentioned, polyethylene, polypropylene, chlorobutyl rubber and, most elaborate but now well known, the polymer of ethylene-propylene diene monomer, fortunately enjoying the acronym EPDM. At an earlier time in this post-1945 chemically creative activity there were fabrics impregnated with 'Neoprene' or 'Hypalon' which are elastomers, but they now seem mainly to be memories. In order to simplify the discussion, reference will concentrate on EPDM and PVC as representatives of the two types.

EPDM is a rubbery polymer. Its colour is normally dark grey due to a high content of protective carbon black. It is a thermoset and therefore cannot be welded; the laps are sealed by a compatible proprietary adhesive to band widths recommended by the maker.

PVC, polyvinyl chloride, is a very adaptable non-rubbery polymer. In its simplest form it is hard and inflexible but it is receptive to an exceptional variety of plasticizers, additives, extension aids, etc. which make it possible to extrude it, e.g. for window frames, or to use it for injection moulding or to calender it into sheet form, as for roofing. It is thermoplastic and therefore can be welded. The surface finish of the calendered sheet for roofing is slightly matt, modestly easing access for light and sometimes giving dust a bit of a foothold.

Laps are made fairly wide, 80–100 mm, hot welded but not necessarily for their full width. One recommended technique tacks the lap at intervals and then uses a continuous weld along the back edge followed by a second one near the front edge. The need for two bands is a reminder of the limits of perfection of operatives doing such jobs on-site.

The rolls of any type of plastic membrane are heavy and if not aligned accurately for unrolling there will be a temptation to do an adjustment to get a constant lap width as the roll is pushed along. Such adjustments can introduce slight wrinkles which can seem initially to be welded flat, but because plastics have memories of their relaxed state and try to recover it, the wrinkles may eventually return and open up by peeling and then they can be difficult to correct. Getting the material laid right in the first place is an important factor in obtaining trouble-free service life but it may not be easy to do and it needs watching by on-site inspections.

The life of polymeric roofings is usually warranted to either 10 or 15 years, which is the time the makers estimate that they will stay free from reasonable complaint in normal service. In fact they may exceed this by as much as 40 or 50 per cent before degradation reaches the point of making replacement necessary. The deterioration will normally show itself both as a loss of the properties needed for roofing service and the development of faults along the laps.

Where roofs may have to be perforated during their life to accommodate changes such as new air-processing equipment or toilet rearrangements, it has sometimes proved difficult to make the synthetic roofings water-tight again. Industry often needs such changes and so do modern shops. Where this is foreseen as a possibility, architects should check that water-tightness can be re-established by the kinds of sub-contractors likely to be involved.

The weathering of plastics

Degradation of plastic materials, apart from plasticizer loss, is an ongoing subject of research to find ways of delaying its development. The risk is mainly damage to the polymer chains and their linkage, and the principal causes are UV penetration, photo-oxidation, water entry and exit, dust, and air-borne pollutants. Protection is usually by additives of various kinds to exclude the alien influences but care in manufacture can also be a significant contributor to longevity.

Reactive air-borne pollutants are typically sulphur dioxide, nitrogen oxide, various hydrocarbons and two of their derivatives, carbon monoxide and carbon dioxide. Some of these are now particularly prevalent in urban areas as products of internal combustion engines, so that some plastics suffer more in urban areas than in rural ones.

A well-known sequence of plastics degradation is the formation of ozone, by reaction between sunlight and atmospheric nitrogen oxide, and the ozone in turn induces micro stress-cracking of the outer surface of some plastic roofing membranes, mainly elastomers. The consequent degradation of the surface by micro-crack propagation gradually eases the entry and exit of moisture, causing cyclic internal stress by wetting-expansion and drying-shrinkage. Loss of surface gloss is an early indicator that micro-degradation of this sort is occurring. It is said[5] that an early indicator of smog effects in California was an increase in the stress cracking of rubber and rubbery plastics. Ozone attack has to be contained in order to delay the loss of tensile strength and puncture resistance of these materials.

Resistance to photo-oxidation is the main aid in extending life expectancy and two types of additive are commonly used for this purpose, pigmentation and metals in ionic form, the first working by obstructing the entry of radiation and the second by its reflection. Used as single-ply membranes, these roofing materials have low thermal capacity and become hot or very hot when laid over insulation, perhaps reaching 70°C or more, and the heat becomes an agent favouring the loss of volatiles and acceleration of other ageing processes. By the same token they also undergo abrupt changes of temperature in variable weather, sometimes dropping several degrees per minute in cold rain for example. This is not helpful to durability because it increases strain rates, reducing accommodation by the material and facilitating crack propagation.

Dust does its damage mainly by harbouring water so that for longish periods the plastic surface may be subject to higher humidity levels than the normal equilibrium moisture content of the material, and sometimes dusts also contain aggressive chemicals or biological agents which can initiate attack or support undesirable micro-growth on warm surfaces.

Apart from the research which is devoted to prolonging the service life of plastics, there has been an evident tendency to increase thicknesses as a contributor to that end. The guarantees offered in respect of life expectancy are usually conditional upon the use of licensed sub-contractors because the quality of workmanship in making laps and edge details is a major factor in achieving a full service life. Sometimes the licensed firms are expected to check over the installation after a period of two or three years with guarantees that then run from the checking date; this makes a lot of sense. If a proposed design is approved by a central office and the work is to be carried out by a licensee, always ascertain which of the two carries responsibility. Some litigation has turned on this.

Quality and liability

The fact that care in manufacture can be important for the quality of plastic products means that it can vary in products of similar description, even those claiming British Standards conformity. Hopefully certification laboratories get it right, but architects and other building professionals should also enquire of colleagues about their experience of unfamiliar products under consideration.

In its first few years, when the material itself has not degraded much, the laps will be the main source of failure risk. The operatives making them will be skilled, but there may be several kilometres to be made on a single job and, as remarked earlier, perfection is not a reasonable expectation on building sites, and especially for roofs. Faults will occur, especially in poor weather or when a contract is running late. Later, beyond the guarantee period, the risks to the material itself will build up as degradation begins to take its toll, and single membranes have no second line of defence.

Making laps is expensive, and this and their workmanship risks seem to have encouraged manufacturers to increase roll widths to reduce lap length, but the rolls are then heavier and less easily shifted about for adjustment of the rolling line (see also page 213).

Fixing membranes

Single-ply plastic membranes laid over insulants are typically fixed by clips, usually comprising a plate with a coating of the same material as the membrane, with a long, special self-tapping screw passing through it down to the deck. The screw goes through the edge of the plastic to be lapped so that the overlay lap can be sealed over the plate. The screw shank has to go through both the insulant and any vapour barrier used in order to engage the deck material and, of course, this can threaten the barrier integrity.

If the deck is of wood or a wood product and gets wet during construction or later from condensate, steel screws may be destroyed by corrosion even if they have been galvanized. The roof covering will then have lost its integrity and will be vulnerable to removal by wind lift.

Some plastic membranes can conveniently be fixed to the insulants by adhesive and it may be sensible to use them when the deck beneath the insulation is concrete, because fixing reliably into concrete by screw clips is difficult. If adhesive is to be used, it should be done by the licensee to the maker's recommendations using the relevant proprietary adhesive, and the insulant must have a suitably smooth surface to receive it. The adhesive will not be bituminous because bitumens and the sheet plastics used for roofing are not compatible. An intriguing case had to be investigated.

Case note

The profiled metal roof deck of an air-conditioned factory was overlaid by insulation and a dry-fixed PVC weather membrane, held in place by metal clips self-tapped through the insulation into the metal. The management soon noticed that the screw heads were working up through the PVC, allowing rain into the insulant and through to the deck. The conditioned air was dry enough to have been judged not to need a vapour barrier (Figure 9.26).

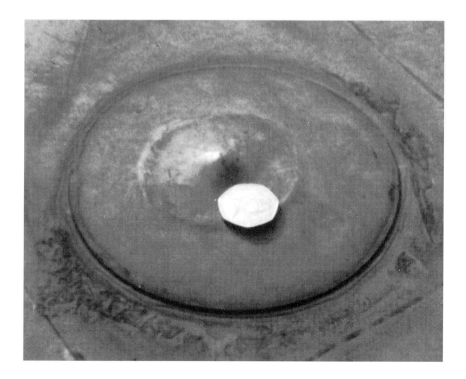

Figure 9.26
A screwhead about to come up through the plastic.

Figure 9.27

But why were the screws undoing themselves? It became evident in the investigation that the pressure of the conditioned air indoors was getting through the insulation layer and pushing the membrane up, keeping an upward pressure on the screws (Figure 9.27); but wind could ripple the unadhered PVC and make the uplift intermittent so that the beginning of each wave gave a little twist to the screws. This seemed to offer a plausible explanation for the undoing of the screws, so a test was made by putting bent drinking straws on the screw tips where they poked through the underside of the profiled metal. The rotation was found to be taking place sometimes as fast as 90° per week.

The case is described partly because of the amusingly simple test technique; but screws that stay fastened have now been developed and should be used.

Fixing by ballasting is an option and is usually done with rounded gravel or larger pebbles and has the advantage of protecting the membrane from UV irradiation. It needs to be at least 50 mm deep or its equivalent weight in pebbles to avoid uplift in wind.

Limitation of pedestrian access

Single membranes overlying insulation are not suitable for free pedestrian access because of puncture risks. The makers of the materials usually have their own preferred recommendations for walkways for maintenance and other necessary access and it is sensible to use these.

Edge seals

It has already been emphasized several times that the edges of the insulation void have always to be well sealed. The vapour barrier is likely to have to be a plastic ply in order to be compatible with the weather membrane, and makers of the latter can be expected to supply a suitable barrier material.

The details which are published in their trade literature usually show the barrier turned up at the sandwich edge but not always, if ever, sealed properly to the weather membrane, perhaps because it is not easy to do. The seal is as necessary as it is with bituminous membranes if the void and the insulant are to be kept dry, and it must be made.

9.2.3   The protected membrane arrangement

The problems discussed thus far mainly arise in one way or another from placing the main roof insulation somewhere beneath the weather membrane, either below the deck, forming the deck, or above it in the sandwich arrangement. Manufacturers have worked hard to overcome the difficulties this creates but these often challenge the laws of physics, chemistry and mechanics. The protected membrane arrangement makes possible the avoidance of most of the inherent conflicts. In principle, this places the weather membrane directly upon the deck and then protects it by the insulation, laid dry and kept in place by ballast or paving. There is no separate need for a vapour barrier (see also pages 191–2).

The insulation is accessible to rain and must therefore be effectively waterproof. At the time of writing the one best known for the purpose is the familiar extruded polystyrene. It is vulnerable to degradation in sunshine and one of the functions of the ballast is to keep it shaded; when thus protected it lasts very well. I have lifted examples, in one case seven years old and in another more than twenty, looking unchanged.

As noted earlier, the material does undergo some shrinkage but in the protected membrane position it is laid dry and so far as I know its shrinkage has only once caused a problem. It was with the weaker type of asphalt and it seems that under the weight of the ballast the polystyrene may have developed just sufficient frictional hold on it to start pulling it apart along the joint lines of the polystyrene when its shrinkage took place (Figure 9.28). Normally there appear to be no difficulties.

Foamed glass blocks have become available in a form suitable for the protected membrane arrangement but I have no experience of their use.

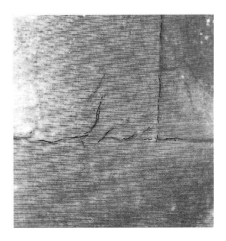

Figure 9.28

Asphalt and bituminous felts

These membrane materials can both be expected to have a very long life in the protected arrangement, upwards probably of 60 years for good asphalt and perhaps 50 or longer for good bituminous felts. There appears to be no chemical incompatibility between extruded polystyrene and bituminous materials.

As regards bitumen felts a question does arise as to whether to use one layer or more than one. A high-performance felt bedded in bitumen should have a long life in conditions free from photo-oxidation and tensile stress, providing that laps have been made well and that jointed decking does not become restless and open up its joints. There seems to be no reason to suppose that the latter will occur in this protected position because conditions should be thermally and moisturally stable, or reasonably so. If a risk is foreseen of deck joint movement, the felt should be only spot-bonded to the substrate as described earlier. Laps in the felts will benefit from the constant pressure on them of the weight above them. Blistering is not a known risk in this situation.

The habit of using bitumen felt in several layers is so well established that the idea of using it as a single ply seems almost to suggest wilful risk, yet it is exactly paralleled by single plastic plies used in this way. However, assurance can be made doubly sure by using a cap sheet on the felt.

Plasticizer loss

Single ply plastic membranes which contain plasticizers have two particular risks in the protected membrane position. It will be recalled that the plasticizers may be squeezed away from positions where pressure is applied (page 212), and there has been a trade recommendation that PVC should not be laid direct on the roof deck but that a so-called fleece – a thin sheet of foamed polyethylene – should be interposed to avoid damage to the membrane by any roughness on the deck. PVC is in fact a very tough material, unlikely to be mechanically damaged by such roughness, and it seems possible therefore that the risk foreseen by the makers is that roughness could create pressure points causing local loss of plasticizer, leading to some perforation. That is one risk.

The other is that a recommendation has also been made for a fleece to be placed between the PVC and the polystyrene on the argument that bits of sand or grit from the ballast can work their way down between the polystyrene sheets to the PVC and may cut through it under the pressure of the overlying insulation and ballast. Pin-hole perforations have certainly been found in PVC in this protected position but, as just remarked, the material is really very tough and it seems that local loss of plasticizer due to bits of grit under pressure is again a more likely cause than any cutting action.

Another reason for a separating fleece can also be suggested. Given the opportunity of direct contact, plasticizer migration would take place from the membrane to the polystyrene and the fleece would act as a separator to prevent this. Certainly there have been hearsay reports of PVC membranes being found drum-tight under polystyrene without the fleece, suggesting a loss of plasticizer, but if the fleece is so important in this upper position, what has then to be borne in mind is whether maintenance staff or operatives doing modifications to such a roof in the future will always know why the fleece is there and replace it correctly.

Non-migratory plasticizers are now sometimes being used. It is evidently important to establish whether a plastic roof membrane being considered for use contains a migratory plasticizer.

When extruded polystyrene was initially marketed, it had square edges but subsequently these have become stepped. One reason put forward was that it would help to prevent sand or grit in the ballast from working its way down to the membrane. With asphalt or bitumen felt membranes it would not matter, and the plasticizer risks for plastics have been explained. It is difficult to see any good reason for the step which is otherwise a modest nuisance on-site.

It has sometimes also been suggested that a significant loss of insulation value occurs due to rain getting under the insulant and carrying away some building heat as it goes down the drain. This has to be nonsense. Measurable precipitation in Britain ranges on average only between 5 per cent of time in the driest areas of the country and 15 per cent in the wettest parts. Much of this never becomes running water but evaporates as surface moisture on the ballast or the insulant, so the total of precipitation to be drained away in liquid form occurs only in a very small proportion of time. Its significance is further diminished by the fact that it can only matter in cold weather and would be beneficial when it is hot. Finally, very little rain gets under the insulant anyway and the small amount that gets there simply stays there. The illogicality of the suggestion is such that one wonders why it was ever put forward. It cannot possibly offset the very substantial advantages of the protected membrane arrangement in other respects.

There is a minor but useful point of technique to note. The insulant can be laid to span gutters (Figure 9.29) and in this way keep them largely free of coarse debris, leaves and so on. Rain gets through easily.

The recommended ballast is rounded gravel but crushed stone is not uncommon in North America and would avoid using the increasingly scarce gravel resources of the UK. The ballast must be larger than that which can get through drainage gratings but small enough to ensure good shading of a sun-sensitive insulant like polystyrene. Paving slabs are equally good and Figure 9.30 shows a very handsome roof combining gravel selected for colour and pavers for maintenance access.

A useful extension of the paving concept is to use the gravel as a base for reinforced concrete patio and leisure platforms cast *in situ* on it. Examples exist of such areas cast about 75 mm thick and up to 3 m square, with light reinforcement in stainless steel. It illustrates the versatility of this form of roofing (Figure 9.31).

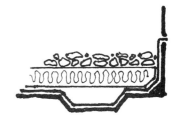

Figure 9.29

Figure 9.30
(Courtesy of the Wimpey Laboratory.)

Figure 9.31

Seeking leaks

There is always the risk with any flat roof that a leak may occur and that a search will have to be made to find its cause. On sandwich roofs this is inevitably destructive but protected membranes offer non-destructive access to the risk positions.

Roof gardening

Roof gardens are *de facto* protected membrane systems. The technology is a specialized aspect of landscape design and will not be discussed here beyond saying that it is usual to lay a filter layer of garden material as a base and then a permeable membrane of some kind over which the top soil is then laid.

Old-established roof gardens which have subsequently been lifted have demonstrated the longevity of asphalt under such protection and the soil itself gives excellent insulation. A useful reference is available.[6]

Testing

It is usual and sensible to test the membrane for leakage before laying the insulation and ballast or the garden soil by plugging rainwater outlets and flooding the roof.

The protected membrane arrangement is logically sound in principle and detail. The deck is protected against thermal movement and the membrane is protected against environmental degradation and mechanical damage. The appearance can be good, the durability high, the freedom from incidental problems excellent, and exploitation for roof usage easy.

9.2.4   Sheet metal for flat roofs

Sheet metal is not often used as a finish for flat roofs but it has been done occasionally and several misfortunes have been investigated. A key point is that sheet metal needs a hard deck for support and this implies the use of a cold roof and with it the likelihood that condensate will form between the metal and the deck. Corrosion is then probable. Two types of risk follow from this and can be illustrated by a case.

Case note

An aluminium roof was laid on a flaxboard deck with polythene as an interlayer. The metal perforated and leaked. Initially it was assumed that the perforations were due to air-borne pollutants from a nearby industrial plant but it was soon appreciated that they were the result of corrosion from beneath. The flaxboard was then found to be saturated by condensate and leakage and when analysed it was found to have a chloride content of 40 parts per million, which is significant.

It was a recipe for corrosion. Condensate was bound to form under the metal, and deck materials which contain natural acids cause such condensate to become acidic. It is locked into an unavoidably thin interspace between the metal and the decking which is then effectively anaerobic. The acid makes the condensate actively electrolytic and thereby facilitates the formation of the anodes and cathodes by which corrosive perforation develops, and the lack of air deprives the situation of a useful preventative and healing agent. An extended discussion of corrosion processes and prevention is provided in Chapter 10.

9.2.5   Notes for designers

- The right place for insulation is above the weather membrane where it acts protectively. When placed beneath a membrane it is destructive.
- In the protective position it should lead to a life expectancy of half a century or more for bituminous membranes. Roof gardening is in this category.
- When a membrane has to be placed above insulation, the latter must be strong enough in compression to resist damage to the membrane by local loads.
- The composition of asphalt can be crucial to its freedom from misbehaviour.
- Bonding together materials with very different properties always introduces risk.
- Thin pavers applied to bituminous roofs can cause damage to membranes. Concrete pavers can expand collectively and do damage at the perimeter.

Three cases of the unexpected in design

Some pipes or tubes that go through the roof to atmosphere may have to be able to move up and down a little in relation to the deck if they are subject to warming and cooling or if allowance has to be made for deck deflection (Figure 9.32). When there is a cluster of pipes like that in Figure 9.33, it is impossible for operatives to do a water-tight job; there should have been room enough around them for decent workmanship or they should have come up into a common vented housing. It can also

Figure 9.32
No way could this ever have worked.

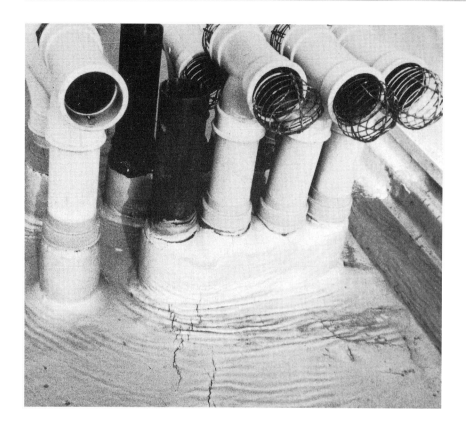

Figure 9.33
An impossible job to make water-tight.

become impossible to do a seal reliably around a pipe if it is too close to an upstand. Such situations should not happen but they do, and they illumine the way the well-informed design-mind ought to be able to work.

An odd point that occasionally arises is that rain water downpipes running through ceiling voids can cause indoor water vapour to condense on them if cold winter rain chills them down to dew-point temperatures. Several cases have been seen and one was of particular interest.

Case note

The reading room of a university library normally held about 200 readers. There was no air conditioning or even powered ventilation, and the room was warmed by a proprietary heated metal pan ceiling, perforated and carrying fibre-glass pads. Humidity often became high and positively tropical in the void.

Complaints of roof leakage were made because water came through the ceiling when it rained. It seemed logical enough, but what was really happening was that the roof drainage pipes ran horizontally in the ceiling void and that the air in the void got hot and very humid when the room was full of readers, so when it rained, the run-off instantly chilled the pipes and caused condensate to form on them, which then dripped down along two or three metres of the pipe runs. No leaks existed.

A postscript for flat roofs

Flat roofs were the greatest sufferers in the intensive learning period of the 1970s and 1980s. They are the most severely exposed elements on buildings, and when insulation standards had to be raised, the exposed weather membranes became subject to thermal stressing for which they had never been engineered. Failures were widespread, but they gradually taught us performance differences between asphalts of different composition, they led the makers of roofing felts to produce higher-performance products, and they demonstrated the value of insulants which could be used protectively over membranes.

But it took some twenty years for this learning process to run its course, and the costs have been extremely high in economic terms, in emotional strain for many people, and in reputations. It is the clearest example from this period that the building professions, the industrial side and research together must find a better and quicker way of recognizing and correcting situations of this kind. They are certain to recur.

9.3 Edge conditions

9.3.1 Initial comments

Flat roofs mainly encounter two types of edge condition: upstands such as parapets, plant housings, and the walls of higher parts of the same building, and low-rise edging where roofs simply cap or oversail the lower walls. These will be discussed mostly in relation to roofs having bituminous membranes – asphalt or built-up felting – because design decisions about these are in the hands of the building designer. Plastic membranes are generally marketed as proprietary systems which include edge components, and if the designer opts for plastics it is usually wise to use the proprietary components. These will embody the relevant chemical technology and should have benefited from the maker's site experience with them. This approach will also ensure that liability lies definitively with the proprietor or licensee.

There will nevertheless be situations for which proprietary systems do not provide a complete installation and the published details in the trade literature draw attention to these by referring to items 'to be provided by others'. Design decisions will then be needed that lie outside proprietorial liability and some comments must be offered about them.

9.3.2 Parapets, walls and raised edges

Of the several sorts of upstand boundaries, parapets have the most exacting problems and can serve as representative of upstands generally. Parapets are usually the upward extension of walls emerging into the open where they are then exposed to the weather on both sides and along the top instead of on one side only, as are the walls. They are subject therefore to more severe wetting and drying than walls and to more severe temperature regimes. It is usual for bituminous weather membranes to be turned up to form a skirting against them which is connected to a DPC through the parapet. This effectively extends the weather-proofing of the roof to the outer face of the building. The principle is important (Figure 9.34).

Figure 9.34

Brickwork upstands

Looking first at ordinary brickwork parapets, there is obviously no need for a cavity in an upstand which is exposed to the weather on both sides, and advantages of simplicity, stability, and consolidation of thermal mass are derived from making them solid above the DPC. They hold together better in what are often testing exposure conditions.

The thermo-moisture movement of brickwork in parapets will exceed that of walls because of its greater exposure. It is most often seen on cavity parapets where it often causes trouble at corners because the intersecting parapets expand along their DPCs and vee outwards (Figure 9.35). It damages the corner brickwork, looks ugly, and usually leaks. It should be avoided by soft joints down to DPC level on each parapet far enough back to ensure the stability of the corner but not so far as to expand significantly itself. A distance of four or five bricks is a useful guide.

Brick parapets need soft joints at intervals in their length in the same way as walls but at smaller intervals to take account of their greater thermo-moisture movement. Intervals of no more than 3 m seem sensible for most brickwork. The joints should have a full layer of some durable soft filler to prevent gap-bridging by mortar. They should be closed off in the usual way with a mastic seal for tidiness and to keep bugs out.

Parapets are sometimes rendered on the back face to exclude rain. Whether this is desirable or not depends on whether the coping is also going to exclude rain. If the coping is going to let it through (see below) it is probably better not to render, in order to allow subsequent drying to both sides of the parapet. However, if the coping is going to prevent entry along the top it is sensible to prevent it from the back face as well, so that wetting and drying is from one side only, as in a wall.

Figure 9.35
Corner damage when parapets expand.

- Bricks containing risk levels of sulphates (page 130) should never be used behind renderings on parapets and can be risky in other parapet positions if they cannot dry easily.

Copings

Usually copings are of bricks-on-edge, pre-cast concrete, or metal. With brick-on-edge copings the question arises whether or not to use a DPC under them. The joints are all perpends and therefore seldom tight, so rain gets in freely. If a DPC is used immediately beneath them, the coping will then alternate between extremes of wet and dry to the extent that it can soon become loose and then sometimes dangerous. If a DPC is moved two or three courses down there will be more stability for them and for the coping, but there will still be some restlessness. If a DPC is not inserted, stability will be even better for the coping, but the parapet will get wetter. There is no perfect solution unless water-tight copings are used. Figure 9.36 shows a coping of expanding brickwork on edge which damaged the entire parapet. If a lime-rich mortar is used for the brickwork, an appreciable amount of white calcium carbonate staining may occur (see Figure 4.1(b)) as the lime leaches out at the DPC and elsewhere.

Historic buildings occasionally have continuous brick copings on long continuous parapets, and these generally behaved well. Why? The answer is probably that low moisture expansion bricks set in lime or other weak mortar were used; the bricks could then expand individually and little collective movement would get built up. When situations

Figure 9.36

of this sort occur and repairs have to be made, the mortar should be closely analysed. It is usually sensible to replicate what exists if it has proved to be reasonably durable. In conservation, always try to stay with success.

Pre-cast concrete or stone copings are a different problem. These go so far towards full protection that it is usually sensible to go all the way and make the joints effectively water-tight. Mortar will not do this because there is so much thermo-moisture movement in the copings that the joints get loose and leak. The coping units should be bedded individually and given mastic-sealed soft joints to minimize leakage and the build-up of collective thermal movement. The units should not span across soft joints in the main parapet brickwork but be related to them. With the top of the parapet thus protected, it is then sensible to render the back face of the parapet.

Although solid parapets usually make good sense, there are circumstances where cavity parapets may be logical and there is then a small but important point to watch.

Case note

An unsupported DPC was laid beneath concrete copings on cavity parapets. The weight of the bedding mortar and copings squeezed the DPC down so that it formed a shallow gutter, and this also drew in its outer edges. The copings had mortar joints that soon leaked and the water found its way along the DPC 'gutter' to laps. Because these were unsealed, the water dripped through into the cavity insulation so that this got saturated and wetted the inner leaf of the walls, resulting in massive indoor staining (Figures 9.37 and 9.38).

Metal copings, usually in sheet aluminium, enjoyed some popularity for a time, but their appearance often left much to be desired. Joints were flat-welted or pop-riveted so that lengthwise expansion and shrinkage was cumulative and severe. Jamming sometimes developed at fixing clips so that shapeliness suffered.

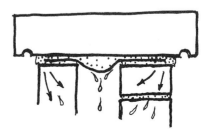

Figure 9.37

Figure 9.38

The logic of dividing up a metal coping into short separate lengths like those of pre-castings lends itself to aluminium extrusions rather than to sheet metal, and in this more rigid form it becomes practicable also to do sensible waterproof joints which can easily accommodate the limited thermal movement. Proprietary systems are available (Figure 9.39).

Figure 9.39
Extruded rubbery strip under the joint.

The parapet DPC connection to the roofing

The DPC which extends the weather membrane through the wall must have a water-tight connection between it and the skirting. In practice a felt-to-felt connection is no problem along the lines shown on Figure 9.34, but it is not easy to make with asphalt and often allows rain to get down behind the skirting into the inner wall system and to run beneath the weather membrane into undesirable places across the deck. Sometimes it gets into the insulation void.

With membranes, the long-established practice has been to widen the face of the DPC joint into a groove about 25 mm square in section and to let the DPC come well into it. Later the head of the asphalt skirting would also be squeezed into it with the aim of making a connection. This was uncertain at best, and mortar pointing was usually inserted to help keep water out and to tidy up the joint (Figure 9.40). In real life, the asphalt often slumps away out of its uncertain contact with the DPC and carries the pointing with it. Somewhat perversely, the pointing and asphalt then not only let water get in but are frequently shaped to catch it and funnel it down behind the skirting. In one case investigated the total leakage into an insulation sandwich on a single large roof was estimated to be of the order of 5000 litres.

Before 1939 protection of this joint by a lead flashing had been a tradition on 'good' buildings, the head of the flashing usually being tucked

Figure 9.40
Theory and reality.

Figure 9.41

into the first joint above that of the DPC. It was kept in place by folded lead wedges and the joint was usually pointed with mortar. Figure 9.41 shows a case where the wall DPC did not even meet the skirting. Bad leakage took place.

After 1945 flashings were usually omitted as an economy, but there were so many leakages that they eventually staged a come-back. However, sheet aluminium is cheaper than lead, is never stolen, and can be presumed to look tidier. The tidiness both of lead and aluminium flashings is a problem because of their large thermal movement. The head of a flashing wedged into a joint will keep somewhere near the temperature of the masonry, while the exposed skirt will get either colder or hotter and the differentials cause distortion. Aluminium has the further problem that it is stiffer lengthwise than lead and even the inserted flashing head can shrink and expand sufficiently to loosen itself and its wedges and pointing. Occasionally wind then gets at such flashings and works them out, sometimes leaving them lying loose on the roof.

It is strange that the building industry has failed to evolve sound, buildable solutions for this small but vital detail, but perhaps the high incidence of faults was explained away as individual instances of careless workmanship. What is needed is a little independent development programme or some proprietary investment with the aim of finding a good standard solution. Some thinking along the following lines might be useful.

The groove could be formed as usual and small GRP channels might be made to sit sideways in it. Then the DPC could be laid to come over the channel to await the later arrival of the skirting (Figure 9.42). Asphalt and bitumen both bond well to GRP, so this might work better.

The skirtings themselves must behave properly. The stronger type of asphalt will minimize slumping and should be used for roofing anyway. Its skirting undercoat should be applied thinly and pressed hard with a

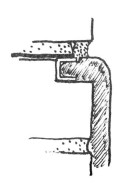

Figure 9.42

small trowel, as recommended in the trade for vertical surfaces, so that it gets well worked into its base. The second coat should also be thin, and applied after warming the first coat by poulticing; the two together should total only about 10–12 mm in thickness. Slumping should then be unlikely.

A built-up felt skirting needs a smooth base for good adhesion. A single render coat should be enough, and the felt needs to be particularly well adhered in case some tensile stress in the main membrane tries to pull it away.

With asphalt roofing, flat areas and skirtings are separate operations and a good bond must be made between them. A butted 90° connection would not be reliable, and a 45° fillet with a face of 50 mm is recommended by the trade and is sensible.

In order to achieve a true bond, however, the asphalt of these elements must be pre-heated by poulticing (page 200) and any sand on the membrane surface must be cut away to receive the fillet. In one case seen, failure to cut the sand away caused intermittent leakage along almost a mile of skirtings – devastating to repair.

The fillet is usually done in two coats as a presumed safety measure. However, both can cool quickly and must be pre-heated by poulticing if the bond is to be trusted. Alternatively one should be done immediately after the other in short lengths to exploit the persisting warmth. Some operatives believe the two are best done as one operation, not separately.

The same principles apply with protected membrane roofing. The fillet connection can then have the advantage of thermal protection similar to that of the membrane, as well as being on a base of the best firmness, the deck itself.

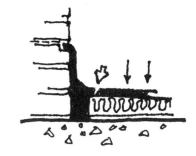

Figure 9.43

A recommendation that is sometimes made when the roof insulation is to be provided in a sandwich is to take the skirting asphalt in a solid plug down to the deck, as in Figure 9.43. This assumes that the insulation is compliant and that the detail will allow for some bending of the membrane at the fillet base even if a load is applied close to it. In fact it concentrates bending stress dangerously at this point and also maximizes the thermal contrast between the membrane and the greater solidity of the fillet and its support. The insulation in a sandwich should never be compliant and then the better solution is a normal well-made skirting.

Where the membrane is a felt build-up, it and the skirting would be done together and to the same thickness, but if spot bonding is used under the membrane, it would not be used behind the skirting. A tilt fillet is put in place to support the 45° junction.

Unusual parapet shapes

Brickwork parapets to balconies have been modelled on the outer face; sometimes as part of an architectural concept. An interesting case raised a problem to be watched.

Case note

The brickwork was narrow at the top, thick in the middle, then narrowed towards the bottom. The builder cut his DPC rolls into what he thought were reasonable widths; they proved to be too narrow but he used them anyway and crossed his fingers hopefully. They failed to connect to the balcony skirtings and failed to reach the outer face. Where the parapets changed angle in plan it proved impossible to form water-tight corners. Extensive leakage resulted.

Stone parapets

Occasionally parapets are done as parts of Classical entablatures in stone. One meets them in conservation work and a few designers use them on new buildings. Their high thermal mass, associated with the low coefficients of thermal and moisture movement of stone, and the use of soft masonry mortar usually seems to have given them enough thermal inertia to remain acceptably stable without movement joints. They should be done to good traditional detailing and specification. Modern innovations should be treated with caution – they may not mix well with traditions.

However, Classical parapets were sometimes built without DPCs and leakage has occurred. In new work it is wise to keep to the principle of connecting the roof weathering through to the outer face of the building via a skirting and DPC in some way.

It is particularly important to protect the upper face of any projecting cornice so that water cannot exploit its vertical joints as ways to seep back into the wall by capillarity or to cause leakage stains. There have been several cases where the top face of the cornice has not been protected and rain has gone back into the wall. In Figure 9.44, the staining is indicative of what is happening farther in.

Figure 9.44

Pre-cast concrete parapets

Pre-cast concrete cladding systems have sometimes been designed so that a top line of pre-castings forms a parapet. This presents almost unsolvable problems for skirtings.

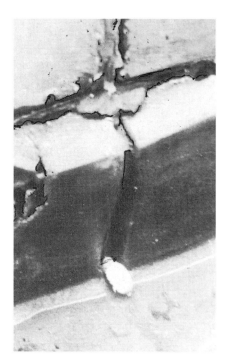

Case note

In one case, the parapet units had a shelf cast on the back to carry them on the edge of the concrete deck, bolted in position after levelling with shims. A groove had been formed along the backs of the units to receive the head of an asphalt skirting moulded in. The joints between castings were nominally 10 mm.

Asphalt cannot cross such a joint reliably. Thermo-moisture movement of the parapet units will split it at the gap, as shown in Figure 9.45 (and confirmed by the little fellow entering what has become his front door).

In fact, leakage was taking place in four ways on this parapet. One was the usual slumping of the asphalt out of the groove with water running down behind; another was the failure of the mastic sealant of the joints to seal to the asphalt – they are mostly incompatible. The third was an occasional shrinkage crack in the pre-castings passing down behind the groove and providing a capillary bypass, and the fourth was general failure of the mastic seals between the exposed parts of the upstanding pre-castings themselves. There is, of course, no DPC in the units which could prevent any of this leakage becoming troublesome and it duly became so.

In view of this difficulty with joints it could seem that matters might be improved by making a pre-cast parapet system effectively continuous. However, this causes another problem.

Figure 9.45

Case note

Parapet pre-castings 3 m long were designed by the engineers to be kept in place by reinforcement knitted into stub columns, making the parapets effectively continuous. No allowance for thermal expansion was made, but it had to take place. One midwinter day the parapet across the end of the building was seen to be out of line (Figure 9.47). It had evidently expanded in the summer and broken its moorings, and although the castings each weighed about 1.5 tonnes, winter winds had apparently been able to work them out of line sufficiently to require their immediate removal.

In concrete-framed systems with attached pre-cast cladding, there is a point to watch about skirting design where the wall of a higher part of the building comes down to a flat roof (Figure 9.46). The cladding designer may be tempted to use the toe of the upper cladding to protect the skirting, but this implies that the skirting must be made before the cladding is attached. Otherwise the position will be inaccessible. This runs counter to any rational construction sequence and also allows no access for repairs. The design of the cladding, the location of any wall DPC, and access for making a skirting and a reliable DPC connection, and the associated assembly sequence, form an important sub-system in any pre-cast cladding solution.

Figure 9.46

Metal-framed parapets

Metal-framed parapets are sometimes used on industrial buildings where steel-framed walls and roofs are used and the deck and the outer

Figure 9.47

wall cladding are both of profiled metal. It is usual then to line the inner face of the parapet with panels of some kind up which the roofing felt can be dressed. Secure fixing of the panels is essential.

Edges for synthetic membranes

Proprietary single-ply membrane systems generally have mechanical attachments for skirtings or adhere the membrane to an independent low-rise upstand close to the parapet or wall. There is then a gap, sometimes quite wide, to be bridged by a flashing which is also required to master the top of the skirting in a weather casting manner rather than by a sealed connection. It is difficult or impossible to do reliably (Figure 9.48).

As remarked earlier, some manufacturers gloss over this kind of problem by saying it is 'to be done by others'. It has to be a metal flashing, which effectively means aluminium, and it has not only got to be fixed in a water-tight manner to the wall or parapet but also has to bridge the gap on a slope so as to shed rain. Aluminium flashings have to be in short lengths for reasons of thermal movement with end-to-end meetings that allow thermal movement and yet keep the water out.

Water-tightness is essential because it is the only protection for the gap, unlike a flashing's normal function as an outer line of defence. Condensate can also occur on the underside of the metal over a gap in this position, and another problem could be vigorous wind activity in vortices behind parapets, which would work to loosen the metal. I have no suggestion to make as to how such a flashing could be successfully

Figure 9.48

contrived nor, judging by some of its literature, am I persuaded that this problem is properly understood by the trade itself. One of the problems with trade literature is that it often seems to have been prepared by the company's marketing division rather than by people who understand what designers need to know in order to make their buildings work.

Gutters

Rain run-off must be able to get away to drainage points, and the higher the rainfall in the area concerned, the quicker it should be able to do so. With protected membrane roofs one could be tempted to do a very simple detail, using the shallow gutter formed where the fall of the roof meets the parapet. The insulation and its ballast could go to the upstand and the rain could filter through and away, but it is surprising how much detritus gathers along the edge of a roof and clogging could soon occur.

A safer solution would be to form a proper gutter and protect it by letting the insulant bridge across it, carrying its ballast as described earlier (Figure 9.29). The gutter would stay relatively clean and the rain run-off would have an easier and more certain access to it.

With the sandwich arrangement it would seem better to take it uninterrupted to the parapet and use the junction as a shallow gutter. One then has a quite customary arrangement where some of the detritus washes away and the rest is cleaned off at intervals.

9.3.3   Low-rise edge design for bituminous roofing

Sometimes designers need a simple, clean meeting line for roof-to-wall junctions. This location is in the highest category of exposure risk and although it has sometimes been detailed successfully, a clean meeting line is technically exacting and has been prone to failure.

Metal trim seemed an obvious option for asphalt membranes. Aluminium extrusions were developed with a horizontal recess to receive the asphalt, a horizontal flange for its fixings and a vertical drip on the outer face. The extrusions were usually supplied in lengths of about 2 m, and the flange (80–100 mm wide) was fixed by nails or screws either to a continuous wood nailer or to wood plugs cast into the deck slab near its edge (Figure 9.49).

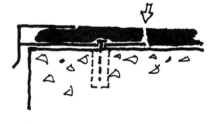

Figure 9.49

It seldom works. The thermal movement of extrusions is too strong to be resisted by nails or screws in wood, and if continuous wood nailers are used they have enough insulation to increase the movement of the metal. The nails or screws loosen in the wood, the flange moves, and a stress line develops in the asphalt along the flange edge so that it splits in long lines. Loosening of the asphalt in the recess takes place at the same time so that the asphalt pulls out a little and water gets in. Fatigue cracking also develops in the asphalt across end-to-end junctions of the extrusions. As a result of all this, water gets under the asphalt and runs down into the building fabric. Also, if the extrusions are set too close end-to-end in cold weather they collide when they get hot in summer and can get broken away (Figure 9.50).

Built-up felt roofs are no better behaved than asphalt in these edgings. Copper, lead and aluminium have all been tried in sheet form for the edging of asphalt membranes and problems have arisen with all of them. The metal typically buckles, often at a welted junction, and lifts the membrane.

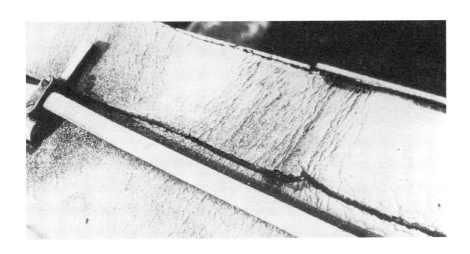

Figure 9.50

A better solution is to develop the edge of the roof as a low-rise parapet using extruded copings in aluminium as for higher parapets. Good end-to-end junctions can be made (page 226), the extrusions can be clipped in place and the clips held in a rain-protected position, and asphalt or built-up roofing can be terminated safely in the shade beneath the extrusions.

An inexpensive option which has been used successfully for asphalt is simply to dress it over the edge of the deck and finish it off with a drip edge as shown in Figure 9.51. Provided strong asphalt is used, it can apparently keep its shape well, at least on a high thermal mass base like concrete.

Figure 9.51

With felted roofs, a similar economy edging can be achieved by a turn-down fold of the cap sheet (Figure 9.52). It has a domestic character and although it may not suggest long-term durability, I have known examples to survive 30 to 40 years.

- All edge trimmings for low-rise boundaries, whether custom-made or proprietary, must be designed to avoid their being lifted in wind and to prevent wind pushing rain up under them.

Figures 9.53 and 9.54 show two cases where optimism overcame good sense.

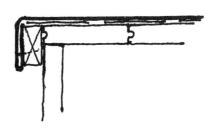

Figure 9.52

Insulation gradients at the roof edge

Bituminous membranes do not like to pass abruptly from one tempera-ture regime to another (page 191–2); they can suffer thermal distress and fatigue splitting. When the sandwich insulation arrangement is used, the weather membrane overlying the insulant is subject to maximal thermal extremes. If, beyond the edge of the sandwich, it is bedded directly on concrete at the edge of the deck, the resulting abrupt contrast can be damaging. The roofing industry has been hesitant about using details which provide a graduated reduction in the insulation at the edge of the sandwich, but something of the kind is necessary.

The protected membrane arrangement is easier and much safer. The membrane is bedded consistently on the deck and benefits from its thermal stability. The insulant can be stopped a little short of the raised edge, and the ballast infill then provides a thermal gradient. If pavers

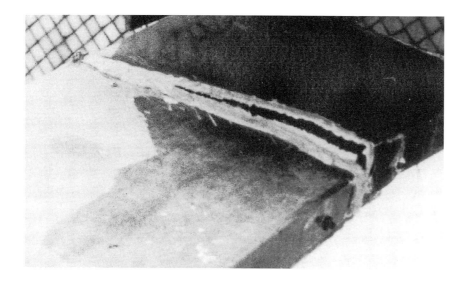

Figure 9.53
A hopeless attempt to seal an end-to-end
sheet aluminium coping with mastic.

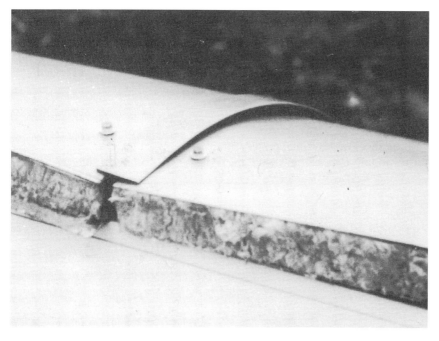

Figure 9.54
An unbelievable attempt to form a curved top
in a building facade.

are laid along the edge to prevent wind-scour of ballast they are an
added safeguard (Figure 9.55).

An unexpected problem has occasionally arisen where very coarse
ballast is laid between the edge-line of pavers and the skirting. The
gradual expansion of the pavers can lock up the ballast so that the expansion pushes it damagingly into the asphalt.

Edgings for single-ply synthetics

The manufacturers of single-ply membranes generally offer shaped low-
rise edgings for mechanical or adhesive membrane attachment. The
shaped form usually provides the low upstand needed to deal with the
risk of water or ballast being blown over the edge. The outer face is
usually a fascia with a weather-cast edge at the bottom. Fixing ('by
others') is invariably to nailers and not always wise.

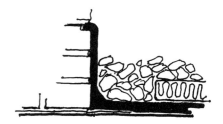

Figure 9.55

Sometimes the shaped section is in sheet aluminium and sometimes it is a hardened form of the membrane material. The latter is probably preferable because aluminium is thermally so restless. End-to-end junctions are not usually shown in the illustrations in the trade literature, almost as if thermal movement does not happen. It does, and it cannot be avoided. Potential suppliers should be closely questioned about the way they accommodate it without damaging their attached membranes. See some examples if possible.

Corners are also not usually illustrated, but cutting, folding or lapping of the membranes has to be done and some membranes do not take kindly to this. Again question the makers closely and get guarantees.

Premature trouble is rare with the main areas of membrane materials; it is the laps, corners and end-to-end junctions of metalwork that are most vulnerable, and attention should be concentrated on them. If in doubt, always ask to see previous installations.

Gutters and drainage outlets

With a low-rise perimeter, the drainage of domestic roofs will normally be by the familiar attached gutters, but with non-domestic buildings some sort of in-board arrangement is usually preferred. When a protected membrane system is used, a solution should be easy along some such lines as those of Figure 9.56, but if a sandwich roof is used, a gutter is likely to be necessary to prevent rain being blown over the edge.

Figure 9.56

Gutters in sheet metal should be reserved for sheet metal roofs, mainly discussed in Chapter 10. They have to accommodate thermal movement by loosely connected weather-cast steps, so they need to be deep, and the membranes used for flat roofs do not bond very reliably to metals, as already noted. Gutters for single-ply plastic membranes should be based on the manufacturers' advice.

Outlets

Drainage outlets down through the deck are similar in principle for all membranes. They usually have a conical base unit comprising a horizontal outer rim and a funnel leading the drainage water into a downpipe. The outer rim is set at the level of the membrane, which is dressed down into the funnel, and finally a mating ring for the funnel may be clamped down on it. There are various ways of fixing the unit to the deck and a grating is set in place to keep out ballast or debris.

The arrangement is simple and obvious but there is a tricky problem to watch. Sheet or felt membranes cannot naturally dress down into a funnel: there has to be a cut and fit operation. Asphalt can be moulded in. Some of the single-ply mastics have a compatible adhesive to help make this junction water-tight, but the bituminous felts may rely mainly on the clamping ring. Of course, the water is already running downward in the funnel and is unlikely to leak back unless there is a blockage, but then it can cause trouble.

It is surprising how often these drainage units are set and sealed poorly, and leak. They should be listed specifically for site inspection. Any single roof should have at least two outlets to avoid the danger of one being blocked.

9.3.4 Movement joints

Underlying most of the thermal thinking in this book is the assumption that insulation will be provided external to the main structural fabric, and it has been pointed out in Chapter 4 that if this is done consistently on walls and roofs, the need for structural movement joints largely or wholly disappears. This offers such great advantages that it is sensible to make this a prime factor in settling the amount of insulation to provide and the detailing to use.

Nevertheless, one may sometimes have to think about movement joints in roofs, perhaps in updating buildings that already have movement joints or for some other special reason. Here, then, are some thoughts about them.

A traditional detail was to provide twin upstands capped with a coping fixed to one upstand only. It works well but is a little clumsy (Figure 9.57). Some makers of plastic membranes have suggested a simple interruption of the deck with a proprietary gap filler, assuming that a sandwich insulant-plus-membrane then crosses it without interruption, but one must then ask whether the movement joint is needed if there is adequate insulation in the sandwich.

What is seldom, if ever, mentioned in the trade literature is how to weather-proof the meeting-point of movement joints in roofs and parapets. There are usually problems but, in principle, it may be easier to use upstands for movement joints when the boundaries are parapets or walls so that roof skirtings can be consistent. A different approach is needed with other edge conditions and it is not practicable to hypothesize about them. They can vary too much. I do emphasize, however, the need to sketch all proposals for these junctions in 3D in advance if they are to be reliably solved on-site in architecturally acceptable ways. Otherwise there will be fudging on-site, claims by builders, and the usual leakage and recrimination.

Figure 9.57

9.3.5 Metal handrails

A small point, but interesting. Cases have been seen where the horizontals of black-painted iron railings expanded in the sun and broke away the concrete into which the uprights were socketed. The horizontals should be sleeved at intervals.

9.3.6 Notes for designers

• Edge treatments for asphalt or bituminous built-up membranes are architects' decisions, but the makers of plastic membranes offer their own solutions as systems. There are then some limitations to think about.
• Parapets are best left solid above roof level.
• They will need to be divided by soft joints more frequently than walls because of their exposure, and corners need protection by soft joints nearby to avoid damage by expansion.
• DPCs in parapets need special thought.
• Should copings be weather-tight or not? (page 224).
• Whenever fresh asphalt is to be joined to some already in place, the latter must be pre-heated by poulticing.

- Pre-cast parapets which are parts of a pre-fab system are almost impossible to weather-proof.
- Most attempts to bond asphalt or built-up roofing to metal edge trim are not successful.
- Avoid abrupt changes of thermal regime for bitumen-based membranes along roof edgings.
- Structural movement joints are a nuisance. External insulation on the roof continuous with insulation of the outer leaf in cavity walls may make them unnecessary.

9.4   Hard paving

9.4.1   Paving materials

Hard paving for roofs, balconies and terraces is conventionally done with concrete pavers, concrete bricks, thin stone, perhaps clay bricks and sometimes clay tiles on screeds. It must always be expected that water will get below any kind of paving. Mortar jointing will not stop it because much of it will not be tight in the first place and thermo-moisture restlessness will soon make the joints loose.

In North America, where pre-cast concrete is usually of a relatively high standard, it is sometimes laid as paving in large units, typically up to 1.5–2.0 m in length and width, with selected rounded aggregate well exposed to give reasonable freedom from slippage in wet or freezing weather.

9.4.2   Bedding

The bedding of pavers is more interesting and important than might be supposed. Sand is usually the main ingredient under concrete slabs, concrete bricks, and stone. Often in Britain it is laced with lime, and under clay bricks or tiles one often finds cement screeds. Lime, cement and screeds all usually spell trouble; on the whole, sand is best left to itself.

The problem is this: lime and cement are alkaline, and when the bedding becomes wet, it stays wet for long periods and the moisture in it becomes alkaline. Some of it will drain away of its own accord but when warmed by sunshine it will expand and be forced out. Some of it will then run to drains but some may be squeezed out along the perimeter and run down the face of the building (Figure 9.58). Even the water that gets under tiling bedded on screed can be forced up on expansion by solar warming (Figure 9.59).

Then, when this alkaline moisture encounters the atmospheric carbon dioxide, it reacts to form a hard, insoluble compound, calcium carbonate, basically white but often darkened by dirt. It can restrict flow through drainage, occasionally building up to complete blockage, and is difficult or impossible to remove from outlets, pavings or wall materials. The sensible bedding is sand alone, trapped in place, or proprietary stools or pads. Mortar jointing should be avoided because it results in collective thermo-moisture expansion.

A paving such as the familiar 50 mm concrete slabs can do this forcefully and cause damage. Clay brick paving has usually followed the now-conventional practice of inserting movement joints, but this allows the bricks to get moved off their individual beds and then ride unevenly

Figure 9.58
Alkaline water running out from sand-lime bedding under deck paving.

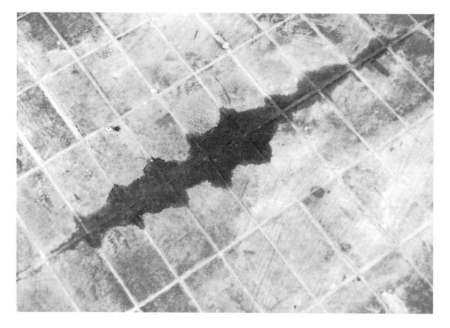

Figure 9.59
Alkaline water expanding under tiling due to solar warming and forced up where it can form disfiguring deposits.

on the resulting roughness, while the soft joints can trip up people wearing high heels or can break umbrella ferrules. A thinner finish such as clay tiles is more likely to arch, sometimes spectacularly, and especially if the main support is a deflecting concrete roof or floor for this will put the central area into compression as it sags.

In continental Europe, where tiling is notably successful as paving, the established practice is to set the tiling in mortar on a sand bed, with the sand locked in place around the boundaries. An experimental area of this kind was laid at BRE in the 1930s and survived heavy laboratory traffic for many years.

The concrete brick paving that has become widespread is bedded dry on trapped sand and appears to be substantially trouble-free. The trade has guidance notes for good practice.

9.4.3 A note for designers

• Avoid using a sand–lime mix or cement screeding under hard paving for flat roofs or balcony decks. Trapped sand is generally the best policy.

References

1 Surman, M., 'Protected membrane roofs', *Construction 31*, pp. 20–21, HMSO.
2 Cullen, W. and Appleton, W., 'The effect of insulation on the durability of a smooth-surface built-up roof', Report No. 7470, National Bureau of Standards, Gaithersburg, Maryland. The Bureau has been renamed The National Institute of Science and Technology (NIST).
3 BS 988: 1973. Specifications for Mastic Asphalt for Buildings; Limestone Aggregate. Replaced in 1988 by BS 6925.
4 Griffin, C. W., *Manual of Built-up Roof Systems*, American Institute of Architects, 1970, in association with US Bureau of Standards (as it then was).
5 Davis, A. and Sims, O. *Weathering of Polymers*, Applied Science Publishers, 1983.
6 *Urban Wildlife Now*, No. 7, *Roof Gardens*, Nature Conservancy Council.

Further reading

1 *Flat Roofing Design and Good Practice*, prepared by Arup R and D in association with a consultative committee and published jointly by the British Flat Roof Council and CIRIA, 1993. A very large compilation of data and general information.
2 Seinfeld, J. H., 'Urban air pollution; state of the science', *Science*, 10 February,1989. 'The urban atmosphere is a giant chemical reactor in which pollutant gases such as hydrocarbons and oxides of nitrogen and sulphur react under the influence of sunlight to create a variety of products, including ozone . . .'

10 Pitched roofs

10.1 Concepts

Most pitched roofs in the UK have relied on lapped claddings to shed the rain and on a loose fit that allows ventilation air to get between and under them to dry off their undersides and supports (Figures 10.1–10.3).

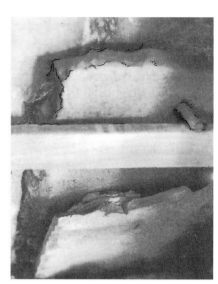

Figure 10.1
The underside of a thirteenth-century stone roof showing peg holders.

Figure 10.2
Traditional slate roof.

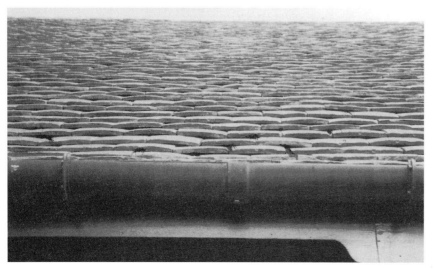

Figure 10.3
Modern clay tiling showing curvature giving ventilation.

Tiles and slates are typical and so, in a way, is thatch. Profiled metals on slopes also rely on laps to shed rain but not to provide ventilation. Roofing felts on pitched roofs are continuous membranes and so are single-ply synthetics.

Two representative scenarios will be addressed: one where roofs are directly exposed to occupied interior space able to be an immediate source of water vapour, and the other where there is a ventilated void between the two to dilute or remove the vapour. The distinction affects the location of insulation. Tiles and slates will be discussed first, then metal, and finally the continuous membrane roofs.

10.2 Tiles and slates

Conventionally, tiles and slates in the UK are fixed to or hooked over timber battens spanning between rafters and beneath which is stretched a thin water-shedding membrane known as sarking felt. Its functions are to offer some wind resistance and to drain off any rain or snow that does get through these loose-fit weatherings (Figure 10.4). Where roofs of this kind are exposed indoors it is usually in roof lofts of domestic property.

All roofs depending on laps get safer against leakage as the pitch increases and the makers of the various claddings have their own recommendations for the lowest safe slopes for their products. The risks are from wind-driven rain, so in places where the Driving Rain Index is high, it is sensible to consider using bigger laps or steeper slopes.

It is always tempting to put the required thermal insulation in the space between the rafters and to support it by plasterboard or hardboard to get a clean underside (Figure 10.5), but this is risky because in cold weather the roofing and battens will get colder and have less ventilation and will stay wetter for longer. This raises the risk of frost damage to the tiles or slates and of rot affecting the battens.

Where there is a loft space with a joisted floor – scenario two and now a common domestic situation – the sensible place for the insulation is between the floor joists with the loft space ventilated. In domestic property this type of roof is often one of the good escape routes for excess indoor vapour pressure and arguably should be kept as such (Chapter 2). The positive ventilation of such a loft should be by a combination of eaves and ridge vents and/or openings in gable walls, helped by fortuitous exfiltration through the sarking laps and the tile or slate cladding. The Guidance Notes for the Building Regulations provide current recommendations for the sizing of eaves and ridge vents and they will do their job by stack and wind power (Figure 10.6).

Case note

In my own house in Hertfordshire, built in 1947, the loft over the bedrooms was separated from them by 9 mm plasterboard, with a plaster skim coat, carrying 100 mm of fibreglass, with a loose-jointed timber floor. The bedrooms and bathroom are all vented direct to the roof space. Eaves ventilation is fortuitous and there is no special venting at the ridge, but there are louvered openings in two gables. No dampness has occurred in the roof and the house itself has not suffered from condensation.

Where a pitched roof forms part of the enclosure of a room in a loft, arguably the safest policy for insulation will be to use it sparingly

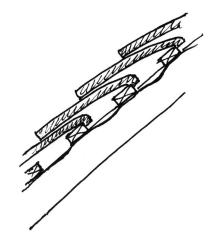

Figure 10.4

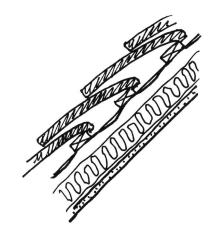

Figure 10.5
Unwise.

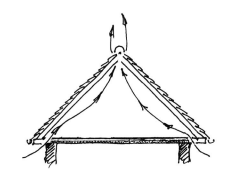

Figure 10.6

between the rafters and leave a ventilated air space of at least 50 mm between tiling and insulation, and then fix a lining board, such as plasterboard to the inner face of the rafters as part of the room finish.

Roof tiles, whether of clay or cement-based, do not normally suffer much in Britain; they are made for this climate and do well in it, but a point about slates needs to be made. Experience has identified which British slates survive well here but some slates from other countries, trouble-free perhaps in their home climate, have degraded soon in Britain, and some artificial slates made here have curled and cracked. Track records need to be checked before specification.

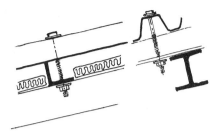

Figure 10.7

10.3   Profiled metal roofs

Profiled metal sheets rely upon laps at the sides and ends to shed rain. The length of individual sheets may be as much as 7 or 8 m, so there is appreciable thermal movement to be accommodated, and the trade practice has been to use laps of 150 mm with a mastic seal. Failures of the seal are more-or-less customary. The slope for profiled metal is typically in the region of 5–10°.

The profiling provides the depth necessary to carry reasonable loads over limited spans in order to accommodate maintenance work. Proprietary systems using either aluminium or stainless or protected steel are available and the protection on steel is offered in a range of colours. In some systems an insulant is placed beneath the metalwork in flat board form or profiled into it. On the underside there is usually some kind of thin decorative finish sometimes described as a vapour check, though this is too optimistic. The supporting structure is typically a steel main frame carrying purlins to which the profiled metal is bolted through its ridges. The side laps may be pop riveted at intervals. The usual purlins are 2's but sometimes angles (Figure 10.7).

Figure 10.8
Torn and ovalized hole in profiled aluminium roofing.

Inherent problems

The profiled metal naturally undergoes thermal movement, greater in aluminium than in steel, although a dark colour-coat on the steel will make them more comparable, and the profiling gives strength to the thermal movement by preventing buckling. Profiled steel can actually fracture bolts by repetitive thermal movement, and laps cannot be reliably sealed because of it. As regards the underside finish, vapour-tightness is seldom a practical proposition. These can be important factors in performance.

Bolts and bolt holes

Thermal movement will be of the order of 1–2 mm/m and the bolt holes will typically be a millimetre or two larger than the bolt diameter. The differential movement of the sheets as a whole may be as much as 8 or 9 mm, so a trial of strength develops between them and the bolts. Sheet aluminium tends to lose out because the holes tear and ovalize, but with the stronger steel sheet there is less ovalizing of the holes and more damage to the bolts. Sometimes they get bent or, where they pass through a lap, the shearing action of the sheets occasionally breaks them in two. Loosened and broken bolts and ovalized or empty holes all provide opportunities for leakage (Figures 10.8–10.10).

Figure 10.9
Shearing action of profiled steel sheet at bolt positions.

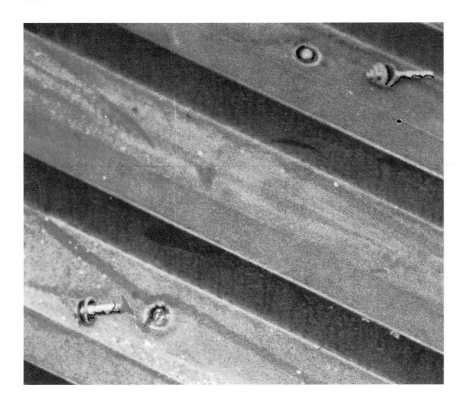

Figure 10.10
Bolts broken by shearing action of profiled steel sheet.

Lap seals

It seems not to be commonly recognized that identically profiled shapes cannot be lapped together in a snug fit; the sides meet first and prevent the tops and bottoms from closing (Figure 10.11).

While it has been commonly assumed that mastic in the laps would be able to seal them, what frequently happens is that the back-and-forth movement of the laps either smears the mastic in a very messy way and destroys the seal or, if there is enough dirt on the metal when the sealant is put in place, the movement rolls it into worm-like shapes that break up, sometimes getting worked out of the lap or being rolled inward. Either way the result is no seal. Naturally the longer the sheet, the quicker and messier is the result.

Mastic tapes were introduced, 2–3 mm thick, a bit stiffer, and tidier to put in place. Their increased stiffness can simply mean that the sides of the profiles engage sooner and leave rather bigger gaps at the tops and bottoms of the profiled shapes than with soft mastic (Figure 10.12(a)). The shearing action can still disrupt the tapes. The effects are obvious if translucent sheets to the same profile are used as rooflights because the disruption is visible (Figure 10.12(b)). The plastic sheets do more thermal movement than the metal and cause more disruption, but the cause is the same and the effect similar.

Although the profiling keeps the main areas of sheeting from buckling, this seems often to occur along the side laps and these are now usually taped and pop-riveted, but what has been seen on some sites is that pop rivets fail by shearing or don't go through both sheets or are simply omitted, and again thermal movement has been seen to disturb the sealant.

A particular problem with aluminium sheeting is that unless it is adequately robust or has unusually frequent supports, the weight of

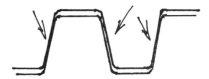

Figure 10.11

(a)

(b)

Figure 10.12
(a) Tape seals broken by movement of profiled plastic roof light. (b) Flattened cheeks of tape seals showing how side contacts prevent full sealing.

peripatetic maintenance personnel at mid-span can bend the ends up by leverage and open the laps permanently (Figure 10.13).

Vapour barriers

It should need only a moment's thought to appreciate the near impossibility of creating vapour seals successfully on the underside of this kind of roofing. The main roof frame, the purlins, the bolts and the carriers for services all frustrate it and in all cases I have seen it has not been successful.

So this type of roof clearly has problems. Some rain may get in at bolt holes or get blown in at opened end-laps or even occasionally through loose side-laps, and condensate can also be expected. Either way, the water may then drop into the insulant and be absorbed to evaporate later or run down the underside of the metal to positions which cause it to

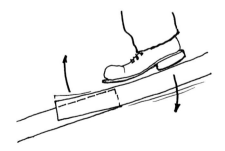

Figure 10.13

drip, for instance at bolts. In one case seen it had apparently then run down the bolts to the steel angles forming the purlins and because these faced up the slope they acted as shallow gutters and held the water until it could evaporate, or until there was enough to spill over at low points (Figure 10.14).

Whether any of this matters significantly in practice depends on the use to which the space under the roof is to be put. If there is not going to be much indoor water vapour and an occasional drip from leaks is of no great consequence, it can provide a handsome and quickly built roof that may be acceptable. If leakage and condensation risks must be avoided, some modifications along the following lines should help:

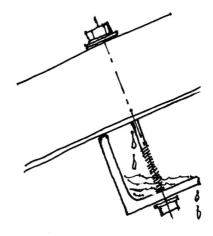

Figure 10.14

- Sheet length should be reduced to a maximum of 2.5–2.7 m or less to reduce the amount of thermal movement.
- Dark colours should be avoided.
- The underside should be ventilated.

Bolt holes should be large enough to accommodate the largest thermal movement of sheets to be expected without cutting into them, and washers should be large enough always to cover the bolt holes well in all bolt positions. Controlled spring-loading of washers should probably be used to allow sheet movement to occur without becoming leaky at the bolt holes, but sheet-coatings may then be scraped. There are practical limits to the precision attainable in fixing because pre-punched holes will be about 2.5 or 3 mm farther apart, or more, when fixed in midsummer than in midwinter, even on sheets as short as 2.5 m.

End laps might much better be left dry but made longer, 250–300 mm, and a big improvement should be possible by blocking the gaps across only the ridges and troughs; the sides will always meet. And the blocking should not be by gunned mastic but done in some way that simply blocks off wind reasonably well. Although an occasional drip may not matter for some uses of space under these roofs, the risk should be discussed in advance with clients and a specific acceptance sought for it.

10.4   Sheet metal roofs

10.4.1   The metals

In the UK the metals most used for roofing are lead, aluminium and copper, with some stainless steel. Zinc, which is popular in northern France and in Germany, has never gained a strong foothold here. The traditional and still-popular way of using metal for pitched roofing is in flat sheet form with more or less continuous support on slopes of almost any pitch, though on one or two of the steepest cathedral roofs lead of the heaviest weight has been known to come adrift and slip after a longish period. The thickness of sheet metals is quoted in millimetres or in code numbers which approximate to the weight in pounds per square foot.

10.4.2   Laying and joining

Aluminium, copper and stainless steel in sheet form are thin and rather springy metals while lead is thicker, as much as 2 mm or more, and ductile. All have quite high thermal movement which must be allowed for in detailing.

Sheet metal is joined either by folding into welts or seams along the meeting edges or by an open roll or a roll formed over a shaped batten. The welts may have a single or a double fold and those that run down the roof slope are usually left upstanding. Cross-welts, when they are used, are flattened to allow rain to run off. Traditionally the folding was done by hand but mostly it is now machine clenched, except with lead.

In making welts the edges of adjoining sheets are turned up at 90° to the plane of the roof with one edge higher than the other, the higher being then folded over the lower. The action of bending the metal lifts the sheet a little on the away-side. The lugs of fixing clips, fixed beforehand to the deck material, are set to come up between the standing edges so that they get folded in with them (Figure 10.15).

Rolls are shaped by hand, and if battens are used they are usually horseshoe-shaped in cross-section (Figure 10.39). Open rolls were common on the traditional lead roofs of cathedrals and churches but battens seem largely to have taken their place, perhaps for better resistance to damage during maintenance. The prominence of rolls and battens makes possible some interesting patterns on roofs.

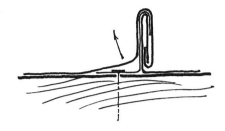

Figure 10.15

10.4.3   Thermal and wind stressing

Mechanical damage by thermal stressing or wind is often either a direct cause of leakage or it may initiate corrosion by the wetting of adjacent wood products in the support system. The design and installation of metal roofing must always allow for reasonable thermal movement, for if this is restricted the metal will crease or pinch and then work-harden at these positions and split. Trouble is normally avoided by limiting the lengths of individual sheets, by designing and fixing them so as not to concentrate resistance to movement, and by free-slipping laps instead of cross-welts whenever practicable. Cases quoted later will sharpen judgements on these matters.

Wind causes damage when it can put negative pressure on the metal or vibrate it and cause loosening and uplift. It happens particularly with the thinner, springy metals, copper and aluminium, and when as a result they are no longer in good contact with their support, the wind increasingly vibrates them so that, again, they work-harden and fracture around or near fixing points or restraints.

10.4.4   Important changes in support techniques

The traditional support for sheet metal roofing was plain-edged softwood boarding, but latterly its place has often been taken by wood products in sheet form, plywood, blockboard, hardboard, fibreboard, flaxboard, chipboard or cement-finished woodwool. A slip sheet of soft, thin, non-adhesive (Erskines) felt is often placed between the metal and its support, although this was never a part of any metal roofing tradition during the centuries when the reputation for durability of lead and copper was taken for granted. Its main value is probably to dampen rain noise.

These changes in the materials used for support, together with innovative roof shapes, have altered significantly the microclimates in metal support systems in ways that have greatly increased corrosion risks. They explain substantially a sharp increase in corrosion failures which has occurred in modern construction, but in order to understand why this should be so and what to do to avoid trouble it is necessary first to understand the corrosion process itself and how it gets initiated.

10.4.5 Common forms of roof metal corrosion

Roofing metal usually corrodes from the underside outward, which is why the microclimate in the support system is so important. There are two common forms of underside corrosion: pitting and uniform attack. The pitting is an electrochemical process, sometimes termed 'galvanic corrosion', operating on small areas of metal and leading to perforation, while the uniform attack is a chemical process in which large areas of metal are eaten away more or less evenly. Let me explain.

Pitting

Metal atoms, like those of all substances, are miniature planetary systems in which a lot of electrons orbit around a nucleus. The electrons have a negative electric charge and the nucleus is positive, and the attraction which negative and positive charges have for one another keeps them together. The total electrical charge in the electrons normally equals that of the nucleus so that the atoms are electrically balanced and neutral.

Electrons can orbit at specific distances from the nucleus determined by the balance of energy and the strength of their mutual attraction, but it is a characteristic of metal atoms that the attraction between the nucleus and some of its outermost electrons is so weak that some of them can get detached by other influences, outside or within the metal, and become free to move around. It is these free electrons which give metals their ability to transmit heat and electricity, two forms of energy which, when imposed on the metal, pass from one to another of these electrons in a flow process. It is also these, incidentally, that give shiny metals their shininess.

The nuclei of atoms which have lost some of their electrons are left with a positive balance of charge. In this state they are restless, ready to recover stability in a newly balanced relationship outside the bulk metal, and because of this mobile readiness for departure they are termed ions (i.e. things that are going, present participle of the Greek verb *eimai*, I go).

However, departure will not take place without both cause and opportunity. Cause will take the form normally of salty or acidic moisture in contact with the metal, for it is among these contaminants that the ionized particles find affinitive substances with which to react, but opportunity can be denied by a protective layer that forms on the metal when it is exposed to atmospheric oxygen and carbon dioxide and which then keeps the ions effectively separated from reactive materials.

If, therefore, this protective layer were to be fault-free everywhere and stay so, the ionized atoms would not leave the metal to combine with the contaminant substances in the moisture and this form of corrosion would not develop. The layer is seldom perfect however; embedded particles of dirt or non-uniform microstructure in the metal, spots of oil or grease and other such blemishes all may cause gaps or weak spots at which reactions can develop. Scrapes, scratches or small cracks opened up by working the metal also create opportunities, but these can heal if air can still get into them. If it cannot then they become risk positions.

In the electrochemistry of corrosion the reactive locations are known as anodes and in battery language they are the positive poles, made so by the availability of the positively charged ions. Activity would be minimal, however, unless released gases and the new substances, the products of corrosion, could move away. For this to happen there must be negatively charged positions to attract them and flow potential between them via the moisture, or they need to be soluble in it.

The negatively charged areas are known as cathodes and these are typically created by gatherings of free electrons and negatively charged ions at positions under the metal surface where, for any of several possible reasons, there is a relatively low electrical potential. In the present context the flow through the moisture takes place by reason of contamination by acids, for these not only provide affinitive ions for combination with the ionized metal atoms but also make the moisture electrolytic, i.e. able to facilitate the flow of this positively charged material from the anodes to the cathodes. The substances formed at the anodes are the products of pitting corrosion and their flow to the cathodes is in fact an electric current. The vigour with which it takes place is a function of the strength of contaminant in the moisture and this will fluctuate by intermittent evaporation and also according to the amounts of acidic vapour or liquid available in the local microclimate.

Chlorides and alkalis occasionally can also make moisture electrolytic but occur less frequently than acids in roof corrosion, though reference has been made on page 221 to one case where chloride ions were found.

The distance between anodes and cathodes is almost irrelevant. It may be tiny, no further perhaps than to the edge of the seat of corrosion (Figure 10.16), or it may be quite a long distance. It matters little; in principle the action is the same.

Figure 10.16
Anodes and cathodes no further apart than the edge of the seat of corrosion (the material peeling away was paint).

The next and most important matter to address is the source of the relevant acids and it is vital to understand this. Mainly they are resident in wood and when this gets damp the vapour or moisture that comes from it is contaminated by the resident acid, and it is also the case that damp wood itself begins a slow decay which releases acidified vapour. In confined, poorly ventilated spaces, which have occurred frequently under modern roofs, this vapour will dampen the underside or back of any metal exposed to them and serve as an electrolyte, and because confined spaces tend to be damp for other reasons, thermal pumping for example, or normal condensation, corrosion becomes a high likelihood. Different woods have different acids, but oak and birch and one or two other woods commonly found in roofs contain acetic acid, which is very reactive with lead.

Plywood has proved to be a particularly troublesome product in the present context for a handful of reasons. First, it has no special ability to resist wetting of the inner material through exposed end grain, but the glue laminae have low permeability and delay drying, so that it can become a slow-release storage system for contaminated vapour, and the thicker it is, the longer will it harbour moisture. Second, the glue layers are commonly phenolic resins containing formic acid, and although the heat of plywood manufacture should lock this up, the analysis of corrosion products in one notable case nevertheless showed a small content of formic acid, which suggests that the locking up is not necessarily perfect. Formic acid is very undesirable in this context. Third, when sheet products are used as direct support for metal there is no prospect of ventilating the interspace so that if the plywood gets damp, the electrolyte it creates is likely to be present in the interspace in a high concentration for long periods.

For sheet products one can read blockboard, hardboard, fibreboard and any other woody sheet material, though chipboard needs a particular comment. Its binder is usually urea formaldehyde, sometimes fortified by melamine and occasionally other additives. This may seem to enclose the wood chips safely, but experience suggests that it would be unwise to assume that the wood cannot get damp or that acidic moisture will not then escape from it.

Flaxboard has been mentioned earlier (page 221) as a source of chlorides, and of course, wood preservatives and the sea are other possible sources of undesirable contaminants. Cement-finished woodwool provides an alkaline surface on which to lay the metal and experience has demonstrated that this is harmless, at least with lead.

There is another way by which anodic positions can be created in sheet metals. Atoms in metals cluster together to form crystals, so that when metal is heavily worked, for example by sharp folding and beating, deformation of the local crystals takes place, and this locks a lot of stress energy into them such that the position becomes potentially anodic, merely awaiting the presence of an opening in the surface metal and an electrolyte to become active as a seat of corrosion. Because heavy working can also cause microcreving of the surface, the corrosion may begin along ragged lines so that eventually the metal may look as if it has been torn (Figure 10.17). The corrosion that is sometimes found near folds for batten rolls or alongside and sometimes inside upstanding welts can have this cause. Sharp folds and snug laps may look like commendable workmanship but, *ipso facto*, may initiate corrosion.

Pitting can take place externally and the most usual sites for such corrosion are in the depths of folds and crevices where airborne dust can

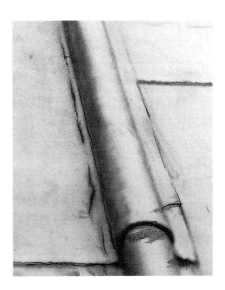

Figure 10.17
Linear corrosion alongside a sharply formed fold.

lodge in quantity. Dust is mainly wind-blown fine sand or earth which will typically contain sodium chloride and nitrate as well as salts of calcium and potassium. Wetted dirt packs closely and is slow to dry. The rain which wets it will be acidic to some degree and perhaps salty if the sea is nearby. If, therefore, there are faults in the protective film on the metal below the dirt level, all the conditions for corrosion will be in place and little air can get at the metal down there to do any healing. Electrolytic pitting is likely to take place.

Uniform attack

Uniform corrosive attack is a more straightforward process, the chemistry of which depends in detail upon the metal being attacked. Lead is quite often afflicted and will be used as our example.

As I have said, some woods used in roofs contain acetic acid which can emerge as a contaminant of airborne vapour or by contact with any other moisture present, and when this deposits itself on lead the initially protective film of lead oxide combines with the acetic acid to produce soluble lead acetate plus some water, and the protective layer no longer exists. The contact air will normally carry some carbon dioxide and this will form carbonic acid in solution on the lead and the lead acetate will then react to form a precipitate of insoluble basic lead carbonate. This reaction liberates acetic acid again and this is free to dissolve more metallic lead so that a continuous process operates for its conversion to basic lead carbonate.

This was discovered a long time ago and was industrialized to produce the white lead which used to be a regular constituent of paint by boiling vinegar under sheets of lead. When dry, the carbonate takes a flocculent form many times the volume of the original lead and an example is shown in Figure 10.18. This process is catalytic and can continue therefore until the lead is consumed.

Figure 10.18
Lead carbonate, 'white lead', resulting from dampness in the plywood and the lack of ventilation.

Nails and screws

Corrosion of nails and screws has been mentioned as a problem in flat roofs (page 215), but they can also come to grief for other reasons more likely to be found on pitched roofs. If they are put into damp wood they can suffer surface damage when screwed in so that anodes develop on the shafts; cathodes may occur anywhere nearby. When buried in the damp wood, no air can get to the shafts to do any healing, the dampness will be acidic and electrolytic, and everything is ready to create corrosion.[1] If the wood contains preservatives or fire retardants, or if it absorbs humidity near the sea, this will be unhelpful too.

Bimetallic corrosion

In buildings bimetallic corrosion is mostly a problem in plumbing; architects seldom use two different metals for roofs on the same building. But roofs need nails or screws for fixings and if these are of a metal wrongly related to the roofing, electrolytic action between them can destroy one or the other.

Bimetallic behaviour depends largely on differences in what is known as the electrode potential of the two metals concerned, metal with a lower potential being usually anodic to one of higher potential. The terminology has a curious class structure about it, the lower potential metal being 'baser' than the higher, which is 'nobler'. Some of the metals which may present problems when used together are noted in the later discussions of individual metals.[2]

Ductility and work-hardening

The crystalline structures which are the substance of metals are ductile in some metals but stiffer and more elastic in others. Ductility depends largely on the size of the crystals. Lead, a naturally ductile metal, has large crystals. Other metals and alloys naturally less ductile can be given some additional ductility by annealing, which is a process of heating and cooling by which crystals can be enlarged.

Intercrystal relations can be disrupted by overworking by operatives or by thermal stress, but even this is a form of work-hardening and it leads to embrittlement and fatigue cracking as well as to anodic situations. A homespun example is the way one bends wire or thin strips of metal to break them, but this takes only a minute or two because the bending is very sharp. Given a few years a similar effect may be produced on a larger scale in roof metal with shallower bending due to thermal buckling or by repeated thermal movement around fixed positions. Examples are discussed later.

The bending that results from thermal buckling between fixings often begins alongside seams and rolls because the temperature of metal on these frequently differs from metal on the flat, and it can also be encouraged when ends of sheets are folded in a cross-welt instead of being allowed free thermal movement. Bulges then provide opportunities for wind to vibrate the elastic metals as described previously.

What may be called thermal oscillation is another fatigue problem. It takes place in the plane of the metal and can happen when it is laid over asphalt or bitumen felt or other sticky material. Adhesion can occur over sticky patches around which thermal movement then takes place and slowly degrades the crystalline structure. This risk is now too well appreciated by metal workers for it to happen often. Work-hardening by these means can fracture metal without any help by corrosion.

This outline of the nature and causes of corrosion is very simplified and mechanistic where the realities are physiochemical in high degrees of complexity, but it should provide a sufficient basis on which to develop the remainder of this discussion.

10.4.6   Individual metals

Lead

There is little to add to what has been said above about lead corrosion. The metal is vulnerable to corrosion by pitting, by uniform attack and by stress, and success depends substantially upon the support system provided for it, which will be discussed later. One matter that can usefully be added here, however, concerns very thin lead fixed to panels. Sometimes it has been used as thin as Code 1, perhaps only a couple of crystals thick, and an instructive case can be mentioned.

Case note

The lead was stuck to chipboard by an elastic adhesive, presumably intended to allow some differential movement, and the panels were on a steep slope. The lead oversailed the long edges of the panels to allow side lapping over battens.

In the event the metal on panels exposed to sunshine developed short, disconnected, mainly horizontal cracks (Figure 10.19). Lead has about three times the thermal

Figure 10.19

movement of chipboard so that in prolonged sunny periods the chipboard might undergo some drying shrinkage while the lead is trying to expand. What apparently happened here was that the thin lead was too weak to expand coherently and instead buckled at intervals and developed fatigue cracks as cyclic heating and cooling took place. Rain was then able to get into the chipboard increasingly easily so that it became wet, swollen and sometimes distorted. The bonding agent in the chipboard was the usual urea formaldehyde, which tends towards alkalinity. It is doubtful if it played any significant role in the breakdown of the lead. Code 3 lead is now the thinnest recommended for any use.

Lead is not greatly at risk from bimetallic corrosion in contact with other metals, but in a marine atmosphere zinc and stainless steels are slightly anodic to it, and it to copper, and lead can seriously affect aluminium. The sodium chloride in marine air apparently reacts with lead oxide to produce a caustic run-off.

Stainless steel

There are three main classes of stainless steel: martensitic, ferritic and austenitic. Each is made up of different alloys which affect various properties, including sensitivity to corrosion. The different formulations are given numbers by the American Iron and Steel Institute which are used widely for specification. The material usually used for sheet metal work on buildings is austenitic. Type No. 316 is currently regarded as the most resistive to corrosion and for this and other properties it is normally recommended for roofs.

Thermal movement of stainless steel is a little less than that of lead, aluminium or zinc, but it is a more elastic and less ductile material and perhaps a little less easy to join by standing seams than by roll capping over battens. It resists work-hardening, it develops a good oxide film if air gets at it freely, and its general durability rating is high, perhaps a hundred years, although there is anecdotal evidence of some crevice corrosion at about 50 years. Its normal thinness for roofing makes it sensitive to denting.

Because of its durability and colour its use as an alternative to lead has been tempting. Lead is a high-value metal and can be stripped and stolen easily from accessible roofs. However, stainless steel has not always been found to be an acceptable substitute because of its shininess and its easy indentation when thin, and these shortcomings, along with the needs of some industrial uses, prompted the development of what is termed 'terne coating', a lead/tin alloy finish giving the steel the matt-grey finish of lead, 'terne' meaning dull.

As currently made in Europe the steel sheet is apparently cleaned by dilute hydrochloric acid and then prepared for coating by passing it through a flux followed by a bath of the lead/tin alloy. It is then rolled to remove excess alloy and an oil which is used in the plating process. This coating is only a few microns thick.

American manufacture is apparently different; the sheet metal is given a tin coating before the alloy is applied, and the coating is some 20 microns thick. Currently there are no UK standards. The material is available from both European and American sources.

Concern has been expressed that the thinner coatings sometimes actually seem to have promoted corrosion of the steel by micropitting. Under a microscope the material can be seen in section to be spongey and it can apparently absorb the moisture needed to form an electrolyte;

then, since the flux has a chloride base, a risk situation can apparently exist.

Corrosion has been seen in one or two cases. An indication that it is happening may be given by brown staining of the coating. The thickness of the 20-micron coating on the American metal and its pre-tinning appear to combine to avoid this shortcoming.

Stainless steel as such has only very modest bimetallic risks but it can be slightly anodic to copper.

Aluminium

Work-hardening is a recognized risk with aluminium and is dealt with by using alloys of differing ductility for different purposes. A distinction is made, for example, between the tempering needed for hand-formed and machine-made welts.

All aluminium is sensitive to chemical attack by alkalis. The thermal movement of aluminium is higher than of stainless steel but lower than of lead.

It is customary to use natural, mill-finished aluminium for roofing. Its protective oxide film is matt-grey in appearance and remains a light colour in clean air but darkens in more polluted atmospheres. Given freedom from corrosion and work-hardening, aluminium roofing has a reasonable life expectancy.

Some types of colouring by paint can be applied but must be done in the factory, although the surface may then suffer if much subsequent work has to be done in putting the metal on the roof. Dark colours, like darkening by pollution, increase thermal movement.

The anodizing of aluminium is a familiar finish and is what its name implies. The whole metal sheet to be coated is made anodic electrically in a bath of dilute sulphuric acid. Then, as in all electrolysis, the water molecules are split into their hydrogen and oxygen components. The aluminium atoms react with the released oxygen to form a relatively thick deposit of aluminium oxide which can be dyed to give the familiar colourful finishes. The film is protective.

In bimetallic combinations, zinc, copper, copper alloys and lead can all be anodic to aluminium, though zinc and aluminium both acquire durable protective films that are said to reduce their bimetallic risk. This may not always be the case and should be checked for individual projects. In marine atmospheres stainless steel can also be troublesome to aluminium, and lead seriously so.

Copper

In respect of copper there is little to add to what has been said generally about corrosion risks, vibration or work-hardening, though there is an interesting observation from the trade to the effect that in working copper it should be given as few blows as possible compared with the frequent, smaller taps typical in working lead. This begs interesting questions about the mechanistic nature of the crystalline structures in metals and the changes that occur in them when the material is worked by beating or is vibrated, e.g. by wind (Figure 10.20). Copper is more sensitive to mechanical stress and needs more stringent conditions for annealing.

The thermal movement of sheet copper lies between that of stainless steel and aluminium and is about half that of lead.

Copper normally acquires a protective green patina of basic copper sulphate but it can take anywhere from five to perhaps twenty years to

Figure 10.20
Copper-clad wall panels. Much of the copper has broken away from its adhesion and has expanded by solar warming and wind vibration. Work-hardening and fractures had occurred.

develop. In its early years externally sheltered copper often stays oxidized, becoming black or dark brown as it dirties. Washings from the copper sulphate are poisonous to many forms of plant life, including lichens, and can stain renderings and stone shades of bluish-green.

Lead-coated copper has had considerable popularity and success for roofing in North America but it is little known in Britain.[3] There are American examples of it over a hundred years old. The coating gradually thins and allows the green patination to take place.

Fixings should generally be by copper or a copper alloy to avoid bimetallic corrosion, but the metal may be attacked by the alkalis in ordinary cements or mortar. Zinc and zinc-coated items are anodic to copper.

Zinc

On exposure to air, zinc acquires a grey patina of zinc carbonate. It is reasonably durable and protective and helps to give the metal a life expectancy of the order of 60–80 years, depending on the severity of atmospheric pollution and the steepness of the roof pitch. Steep roofs cleanse better in rain, and it is notable that steepness characterizes many French zinc roofs.

When zinc is rolled, its grain structure is distorted by elongation of the crystals in the direction of the rolling, and this is said to result in weakness which makes it modestly sensitive to cracking when sharply folded along the grain. However, German and French metalworkers seem to shape it very freely.

Zinc is said to be sacrificial to aluminium and loses its protective film. Copper and copper-rich alloys can also cause bimetallic problems, as can washings from bitumen felt roofs. The bitumen is damaged by UV light and the small fragments that come off have soluble acidity.

10.4.7 Support systems for metal roofs

From what has been said it will be clear that the safe design of metal roofs depends largely on keeping any supporting timber work dry and the underside of the metal adequately ventilated. The way the metal is fixed can also be a factor in success. Designers have not had much help from the available research and literature in developing appropriate techniques but we can further that objective by considering some instructive cases.

Two have already been mentioned in passing, the failure of proprietary aluminium facing on hardboard finished wall panels and similar perforation of a group of aluminium flat roofs laid on flaxboard decking. We will look more closely at these.

Case notes

In the first case the panels comprised a thick moisture-absorbent board faced with hardboard to receive the aluminium, the metal being fixed to the panel by adhesive with the edges of the metal returned around the metal perimeter. The metal apparently loosened itself by thermal movement, rain got in at poorly fixed cills of windows and saturated the hardboard and its backing so that acid-bearing vapour could fill the confined air space between metal and its backing. This doubtless condensed on the metal and pitting corrosion then took place by the usual processes (Figure 10.21).

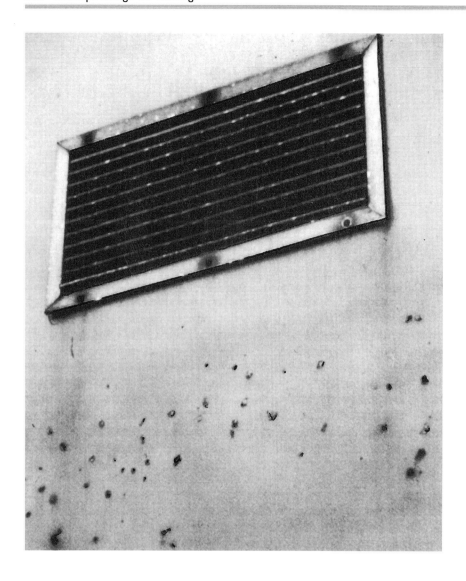

Figure 10.21
Aluminium facing to hardboard faced wall panels wetted by leakage at the vent.

In the second case, the nominally flat aluminium roofs with raised seams were carried on the flaxboard deck, 50 mm thick, with an interlayer of polythene between deck and metal. Along the upper edges of the shallow slopes of the roofs rain had found a way to get blown in and onto the polythene, whence it slowly worked its way over extensive areas of it. Condensate was also doubtless able to form on the underside of the metal. The fixing clips for the aluminium were pinned through the polythene to the flaxboard and no doubt were worked loose by thermal movement of the roof metal so that the flaxboard became saturated and its contaminated dampness infected the moisture on the underside of the metal. The flaxboard was found by analysis to have a significant chloride content and chloride ions were found in the corrosion pits.

To these two examples we can add three of lead corrosion, the first of which is complex but exceptionally important for the present purpose and is therefore described in some detail.

Case note

An elegant free-standing room for meetings was roofed by four lead-covered quadrants of moderate slope separated by a flat cruciform deck of asphalt on concrete (Figure

10.22). Each quadrant sloped inward to gutters along each of the four diagonals, draining to the outer corners.

The lead was laid on 50 mm woodwool panels having a cement finish and the lead sheets were joined over batten rolls. Along the upper rims of the quadrants the batten rolls ended against 150 mm tilt fillets fixed behind the upper edges of plywood fascias 25 mm thick and 0.5 m deep (Figure 10.23), trimmed along the top by a strip of semi-hardwood (Figure 10.24). Wood soffits beneath the fascias were sealed across to the concrete main frame of the roof system against which kerbs were formed. Asphalt from

Figure 10.22
One of the four quadrants; one arm of the flat cruciform deck in the foreground with another beginning on the right.

Figure 10.23
Showing the way the batten rolls ran up to the tilt fillet behind the top edge of the fascia, where the patches cover fractures.

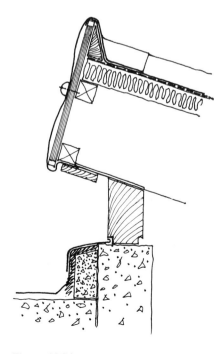

Figure 10.24

Figure 10.25
The fascia lead was dressed over the top of the fascia to master the junction with the roof lead. In this view the rounded top edge of the fascia is densely packed lead carbonate adhering to the peeled back sheet.

Figure 10.26
The remains of a lead welt joining two fascia sheets.

Figure 10.27
Screw fixings of fascia lead through lead dots.

the flat areas was dressed up the kerbs and lead skirtings were dressed over them (Figure 10.24).

The lead of the main roof slopes was dressed up the tilt fillets and pinned down into the top edges of the fascia plywood, while the lead on the fascias themselves was taken up over these edges and dressed down over the heads of the batten rolls and about 200 mm out onto the main roof slopes to provide weathering (Figure 10.25). Along the bottom edges of the fascia plywood the lead was turned back and up behind to be trapped and held by the soffit board.

The fascia lead was done in two lengths along each face, joined by a vertical welt (Figure 10.26), and each length was screw-fixed twice to the plywood by a screw through a solid lead dot (Figure 10.27). Where the four ends of the flat cruciform parts of the roof came to the perimeter of the building, a lead edge was formed over the concrete and sealed against the asphalt by some mastic. The quadrant roofs oversailed the walls of the room by about one full width of lead sheet. The room ceilings below the quadrants were of decorative gapped boarding and carried some fibreglass insulation/sound absorbent, but no effective vapour barrier existed to separate the room air from that in the roof voids, and these latter had no separate ventilation.

This roof is described in detail because it provided so much information about lead behaviour and because so many elements of the design had a role to play in its 20-year progress from completion to failure and replacement.

The first point to make is that thermal movement caused a lot of fractures in the lead which appear to have been a major factor in initiating the failures. In some locations the cracks were of no great importance; a few occurred where the lead skirting was dressed over the asphalt on the kerbs (Figure 10.28) and adhesion to it could have caused local concentrations of stress. Others happened where lead trim was dressed along the perimeter concrete at the four outer ends of the cruciform deck. Evidence of the lead's insistence for thermal movement is clearly visible in vigorous rippling along the ridge-lead of the lower two fascias (Figure 10.29), and dramatically along the fascias themselves (see Figures 10.22 and 10.23).

It is not known whether this lead was particularly susceptible to thermal fracture. Susceptibility is affected by the grain structure and by the amount and distribution of copper in the lead but in practice one has little choice. One uses what is commercially standardized or one uses recast old lead.

However, the thermal damage that mattered vitally occurred in four types of location. Along the ridge over the top edges of the fascias: the longitudinal movement that the lead wanted to make appears to have been restrained by being dressed over the heads of the batten rolls and stress concentration then apparently led to fractures of the ridge metal at these positions. They could then leak rain into the fascia plywood and the tilt fillet (Figure 10.30). They were patched temporarily (see Figure 10.23). The Victorians had a neater way of allowing safely for thermal movement of lead (Figure 10.31).

On the fascia faces (Figure 10.27) where movement of the lead opened up access for rain around the screw fixings: these dots were all loose when examined and the local wetting of the plywood is evident in photos of it after exposure (Figure 10.32). It will be seen that a ring of the white lead carbonate formed around the fixing point and there was intense corrosion of the lead itself here.

The lengthwise movement no doubt also loosened the vertical welts in the fascias a little for rain evidently got in and ran down inside

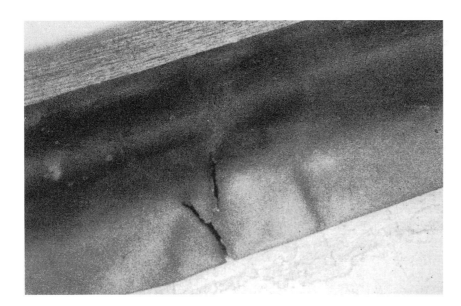

Figure 10.28
Thermal fracture of lead on kerb.

Figure 10.29
Thermal rippling of lead along the back of the top edge of the fascia, and thermal fractures of the flat lead sheet covered by black adhesive felt.

Figure 10.30
Thermal fractures of lead over the junctions of the batten rolls with the upstand of the fascia.

Figure 10.31

Figure 10.32
Concentrations of corrosion around screw fixings and at welt-lines.

them. At the bottom some of it leaked inward and created concentrations of dampness in the fascia plywood and this shows again in Figures 10.18 and 10.32. The wetness appears to have wicked up fanwise in the plywood to produce again heavy concentrations of basic lead carbonate on the metal, while along a sort of high tide mark it will be seen that extremely vigorous linear corrosion took place, showing externally as gaping losses of metal vee-ing downward from the highest tide position behind the welt. This was found behind most of the welts.

It was also the case that the lead sagged a little under the bottom edge of the fascia to form a small gutter along which some of the water from the welt could find its way and get into the plywood by absorption along its bottom edges. When the plywood was checked electronically for moisture content it was found that although it was surface-dry, it was over 60 per cent in the core, an astonishing figure. (See Figure 10.24.)

The only damage to the lead on the main slopes occurred where it was over the eaves. In this position it would be subject to a larger thermal range than on the rest of the slopes and neat thermal fractures took place in straight lines across affected sheets (Figure 10.29).

When the roof was stripped it was more easily seen that the corrosion had its greatest severity around the lead dots, in the downward vees from the vertical welts, within the welts themselves, and along the ridges, but although the main roof metal was destroyed where the fascia lead was dressed over these, the damage ended abruptly where the lead was at rest on the alkaline surface of the cement finish of the woodwool. Here it was in pristine condition. This could affect the choice of material in future for support systems; rendering on stainless steel lath could become attractive. The effect is best seen in the gutter photo (Figure 10.33).

The exceptional exposure of the ridges would no doubt sometimes cause condensate to form inside them and this could explain runnels of

Figure 10.33
The shallow gutter rolled back showing corroded lead over the plywood gutter lining and unharmed lead over the cement finish on the woodwool.

wetted carbonate which ran down the slopes under the lead. The fact that carbonate followed the batten lines closely suggests that the exposure of these also produced some condensate and corrosion.

The four shallow gutters down the quadrant diagonals rested on thin plywood over fibreboard with the gutter flanges dressed out onto the roof slopes where they were lapped by the main roof lead. Figure 10.33 shows that over the plywood basic lead carbonate had formed in a continuous coat while the flanges over the cement finished woodwool again were unharmed. The fibreboard and plywood would certainly have become damp due to condensation on the underside of the lead. The edges of the gutters had already perforated in places.

Some shrinkage cracks in the woodwool had been seen in an inspection a year earlier and where these occurred there were matching thin white deposits. There was no obvious source of acidic vapour nearby and the wide distribution of such gaps argued that the room air itself must have contained relevant contaminants. A test was made by breaking a hole through the woodwool and a year later the corrosion shown in Figure 10.34 had developed. The room is near some kitchens which presumably provided the contaminate. The woodwool itself had become embrittled and was near the end of its safe life for reasons discussed in Chapter 9, (page 186).

Looking back over this description one can see how dampness could take charge in the fascias for it is not difficult to imagine that low winter

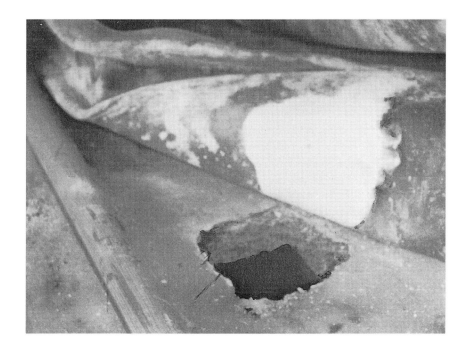

Figure 10.34

temperatures would shrink and eventually fracture the lead wherever it was too greatly restrained and that summer warmth could distil acidic vapour out of deposits in damp timbers, especially in thick plywood, and diffuse it by warmth pressure throughout the connected air channels and spaces.

The replacement roof is similar in appearance but functions in fundamentally different ways. The room below is now sealed off by an effective vapour barrier associated with increased insulation and the void is generously ventilated. The deck now comprises gapped boarding, as do the fascias. At the tops of the fascias a separate lead ridge piece now allows some ventilation beneath it, while the fascia itself is in shorter lengths and drains freely along the bottom. The ridge metal is free to undergo some thermal movement.

In the original quadrant roofs there was no ventilation of the void nor could air changes take place around the fascias. How much ventilation is enough is not known and at present has to be a matter for judgement. The next two cases give some help.

Case note

This was a lead roof on a two-storey building, the lower storey of which was a restaurant. It was a replacement for a lead roof which had failed quite quickly. The earlier roof had been laid on plywood and the replacement was laid on gapped boarding beneath which a 50 mm clear space had been specified over a 50 mm layer of rockwool thermal insulation, and this in turn was laid on a vapour barrier over the plywood of the previous roof. The slope was about 10°.

At the lower end of the slope the lead was turned down for weather protection of the ventilation slot and at the ridge a continuous cowl was formed to protect the upper end of the 50 mm gap. In the event the vapour barrier was found to be loose, partly closing the 50 mm slot along the bottom of the ventilation space, and at the upper end what had been assumed to be a continuous gap was only a series of small-ish openings.

Figure 10.35

One might suppose that stack effect would move air upward through the 50 mm void but it must be borne in mind that in cold weather and at nights the movement would, in theory, reverse. In fact little movement could be detected in either direction.

Some vapour was getting into the 50 mm air space from the restaurant and its kitchens, and where this was happening white deposits were forming on the underside of the lead precisely over the gaps in the boarding (Figure 10.35). How much protection might have been provided had the full 50 mm air layer been able to move freely to atmosphere is not possible to estimate. What seems clear, however, is that such an arrangement must be at best only marginally successful and in practice accident-prone to the point of risk.

Case note

This was another case of a lead roof of about 10° pitch which failed in about 5 years and was replaced, but this time the replacement perforated in less than a year. The replacement had been laid on plywood over insulation over a felt roof acting as a vapour barrier. A slippage sheet of the soft, airy ('Erskine's') felt was placed under the lead and kept it separated from the plywood by 2 or 3 mm. The down-slope ends of the sheet lead were flat-lapped by about 300 mm and the sides of strips were lapped over batten rolls (Figure 10.36).

When the lead was lifted the felt and plywood were found to be very wet and the question arose of how the water got in. It had not previously been assumed that such a thin layer of air could be a thermal pump but it was decided to test it as an hypothesis on the possibility that the good workmanship had so narrowed the laps that they could be sealed by rain or condensate, and that a cold shower

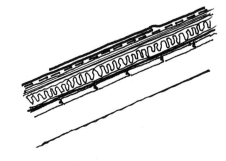

Figure 10.36

Figure 10.37

Figure 10.38

of rain might then chill the air layer in the felted void abruptly and sufficiently to cause a pressure drop that could suck water out of the laps into the interspace. The test showed that this was indeed what could happen, and it was then also realized that because the volume of air was so small and so thinly spread, the abruptness of the pressure drop could probably exert quite a lot of pulling power for at least a short time.

Some of the perforation took place alongside the fold-back of lead where it was shaped to go over the battens and it seemed likely that this was a case where sharp folding had caused crystal distortion which created linear anodic situations (Figure 10.37).

Over the wood cores of the batten rolls the free spaces were sometimes densely packed with basic lead carbonate (Figure 10.38). The batten cores themselves would get damp from the damp plywood, and also, of course, continuous thermal pumping would expel acidic vapour through this escape route in sunshine, so there was plenty of potential for the carbonate to form in these laps.

It seems likely that thermal pumping of the thin air layer could draw damp air in over the batten rolls as well as through the end laps, and the condensate that would then form on the underside of the metal would gradually wet the felt and the batten roll, though perhaps more slowly.

10.4.8 The basics of safe practice

We can now summarize what appear to be the basics of safe practice for sheet metal roofing:

- Never lay it on organic sheet products. They are almost certain to get damp from condensate or in-drawn moisture and they prevent ventilation of the metal face. Use gapped boarding with good ventilation beneath or a cement-finished surface.

- Do not create confined spaces under or behind the metal which could generate thermal pumping.
- Prevent indoor vapour from reaching the metal, and especially if it comes from such sources as kitchens or chemical laboratories.
- Do not put thermal insulation in the immediate support system; it enhances the risk of thermal pumping, it greatly increases the likelihood of condensation on the underside of the metal, and it maximizes thermal stress in the metal.
- Allow metals freedom to make thermal movement. Any unavoidable nailing or screwing must be done where rain cannot get at it.
- It is risky to adhere metal to panels for external use. The two products will have different thermal movements and the metal will probably become detached or be damaged. Some adhesives may promote corrosion.
- All welts, batten rolls and other folded junctions will loosen a little due to thermal movement. Do not use them in ways that let them act as water conveyors, or at least ensure that if they do, they drain freely and harmlessly.
- Remember that lead, in particular, is a very ductile metal which can drift and sag on steep slopes or vertical surfaces.
- Hollow rolls have the advantage over batten rolls that they have no wood in them.
- Avoid creating places externally where dirt can pack and stay damp for long periods.

Are slip sheets worth while? As remarked earlier, it is difficult to see a case for them with lead. The sheet lead on the cement-faced woodwool came to no harm from scuffing and slip sheets were never used on the medieval roofs that had great longevity. There could be more of a case for them under the springy metals, aluminium and copper, to reduce contact noise.

Battens versus open rolls

It has been the habit to take the overlap metal not only around the roll but to dress it out flat for 30 or 40 mm on to the next sheet (Figure 10.37), but this can harbour water in a position where it can distil in sunshine back into the roll. It is now regarded as better practice to stop the overlap as shown in Figure 10.39.

All rolls and seams will heat and cool preferentially because they project from the plane of the roof. Condensate will form in cold weather and subsequent sunshine can distil it both outwards and back into the roll.

Finally, why did the medieval roofs survive so long? On cathedrals they usually had board support that gapped a little by shrinkage, with great voids beneath them fortuitously ventilated. On church roofs they were protected by a policy of ventilating the buildings well. These would all help to avoid corrosive conditions.

Figure 10.39

10.5   Gutters

Metal gutters on the flat or shallow slopes cannot be continuous in long lengths because of the allowance necessary for thermal movement.

Instead the usual practice is to provide loose-fit weather-casting steps at intervals of 2 m or so and at turns and junctions.

Many soldered steps which had broken up at the junctions have been seen in investigations. Wherever soldering is done on metal roofs the flux and the cleaning agent must be removed perfectly along the solder edge if corrosion is not to develop, because flux has a chloride in it. A weak solution of washing soda and a final wash of clean water is normally used to clean and neutralize it. Note it for site inspection.

Quite a lot of depth is needed when gutters have to be stepped and the detail is inevitably a bit clumsy. A safer gutter material could be high-performance roofing felt.

10.6   Noise and sound insulation of metal roofs

All thin metal roofs except those of lead present two noise risks: drumming in rain and easy sound transmission. The high ductility and internal damping of lead discourage both.

Profiled metal roofs in which the profiles are filled by an insulant may have some small advantage by having the insulant in contact with the metal so that it does some vibration damping. The problem is much reduced where pitched roofs are separated from the occupied interior of a building by a loft and an intermediate floor. The roof void allows a lot of the sound energy to dissipate and a floor can easily be designed to provide additional sound reduction.

A more difficult problem is the situation where there is no intermediate floor and some improvement has to be attempted in the sloping roof itself. It has already been explained that where thermal insulation has to be built into the sloping structure it should have some kind of reasonably vapour-tight layer on the inside to minimize condensation. This technique can be exploited to reduce noise transmission by choice of the barrier material. Rendering or plastered plasterboard are possibilities, and a doubled layer of plasterboard glued together to get improved vibration damping can be good. Air-tightness, weight, and internal damping of the lining sheet are the properties to look for.

10.7   Felt and asphalt on pitched roofs

We have come from self-ventilating tiled and slated roofs, through the profiled metal systems which use sealed laps, to the almost sealed sheet metal roofs and arrive now at fully sealed roofs of felt and asphalt on pitched roofs. A pitched roof system of appealing simplicity for domestic-scale construction is a felt overlay on one of the sheet deckings, plywood, blockboard or chipboard. In such use the underside of the decking must be well ventilated; it will not prevent its seasonal moisture movement but it should avoid condensation problems. The usual combination of eaves and ridge vents for the roof void should be provided.

Amounts of seasonal moisture movement depend on the material used. Of the three mentioned, Appendix 1 lists blockboard as having the least movement and chipboard the most, but the chipboard industry has put useful effort into its reduction.

Case note

A three-layer felt roof was laid on chipboard on low-pitched roofs of some terraced houses, each about 9 m long. The chipboard was interrupted for a firebreak at each party wall and the felt was carried across on a slip-board. The roof space was vented only at the eaves. Seasonal movement of the chipboard caused stretching and compression of the felt and eventual rucking and splitting over the party wall line. The seasonal changes in the roof dimensions were monitored by an independent laboratory and were of the order of 17–20 mm per house, i.e. about 2 mm/m. The chipboard concerned was of a type in use in the early 1970s.

There might be some temptation to put insulation between the rafters with a low-permeability inner lining such as plywood so that the roof could be exposed on the underside. Thermal pumping could almost be guaranteed to raise the water vapour content of the enclosed voids and cause trouble.

Asphalt on pitched roofs

On several occasions asphalt failures on pitched roofs have been investigated. In one case woodwool was the base and despite the application of a white finish on the asphalt severe thermal splitting occurred at joints and downslope flow took place (Figure 10.40). In another case the asphalt was on metal lath over boarding containing preservative, and vapour from this softened the asphalt. Its flow bore a close resemblance to that of a glacier and in fact they operate similarly in principle. It is not sensible to use a liquid, even of the viscosity of asphalt, on any slope of more than a very few degrees.

Figure 10.40
Asphalt on a sloping roof of woodwool. The asphalt has a white finish.

References

1   BRE Digest 301: September 1985, Corrosion of Metals by Wood.
2   PD 6489: 1979: A Commentary on Corrosion at Bi-metallic Contacts and its Alleviation; BSI.
3   Fishman, H. B., Darling, B. P. and Wooten, J. R., *Observations on Atmospheric Corrosion of Architectural Copper Work at Yale University*, Special Technical Publication 965, American Society for Testing and Materials (ASTM).

Further reading

1   Bordass, W. T., Dicken, G., and Farrell, D. M., Corrosion control – sheet metal roofs, *Architects' Journal*, 29 November, 1989.
2   Farrell, D. M. and Lowe, R., 'Alternatives to traditional lead roofs', a paper to an English Heritage Conference, December 1994. The authors are members of Rowan Technologies Ltd.

Appendix I
Dimensional changes due to temperature and moisture

BRE published a trilogy of Digests (Nos 227–9) from July to September 1979 describing the changes of size and shape which take place continuously in the building fabric. The key dimensional data were given in the middle Digest, No. 228, but were quoted as percentage linear changes of size. Architects and builders are likely to find it more convenient to have them in millimetres per metre or millimetres per three metres as more direct measures of the dimensional activity which has to be allowed for in detailing buildings, and by agreement they are printed below in this form. Additionally this appendix incorporates some observations either from the Digests or from BAP experience.

The thermal data are accompanied by the temperature ranges to which buildings in Britain are subject, again taken from the Digests; they are in some ways quite dramatic in themselves. The material was first published in this form in the *Architects' Journal*.[1]

Table A1 Thermal movement for external exposure

Material	Seasonal temperature range (°C)	Movement mm/m	(mm/3m)
Cement-based composites			
Mortar and fine concrete:			
Dark	70–85°C	1.1	3.3
Light	but see note below	0.9	2.7
Concrete:			
Gravel or rock aggregate		1.2	3.6
Limestone		0.6	1.8
Block and brickwork		1.0	3.0
Aerated concrete		0.7	2.0
Brickwork (clay, sand-lime) and tiling			
Clay brickwork	70–85	0.7	2.0
Sand-lime brickwork	70–85	1.2	3.6
Tiling	85	0.5	1.5
Natural stone			
Granite	85	0.8	2.5
Marble	70–85	(max)	(max)
Light colour		0.4	1.2
Dark		0.5	1.5
Limestone	70	0.8	2.5
Slate	85	0.9	2.8

Note: all these data are for conditions at or close to the exposed surface of the materials. The average temperatures within the main body will be less, but how much less will depend on their mass and how easily escaping indoor heat can reach them, as well as their thermal conductivity. If they have substantial mass and heat can reach them easily, the 'body' temperature range may be as low as a little over half of the figures shown, while if insulation keeps indoor heat from reaching them and/or they are not very substantial, the temperature range can get close to the quoted figures. Dark surface colours will cause larger movement than shown.

Metals			
Structural steel (see also preceding note) Metal in sheet form	70–90	1.0	3.0
Stainless steel	85		
Austenitic		1.5	4.5
Ferritic		1.0	3.0
Aluminium	85	2.0	6.0
Copper	105	1.8	5.4
Zinc	105		
Parallel to rolling		3.5	10.5
Across the rolling		2.4	7.2
Lead	105	3.3	10.0
Glass			
Clear	65	0.7	2.0
Solar control	115	1.3	4.0
Asphalts and plastics		approx.	approx.
Asphalt	85–105		
Thermoset plastic	85–105	8.0	25.0
Phenol and melamine formaldehyde		2.8	8.4
Urea formaldehyde		4.7	14.0
GRP	85–105	3.7	11.0
Wood (soft and hard)			
With grain	105	0.2–	0.6–
Across grain	105	0.6	1.8
		1 mm in 13 cm	1 mm in 13 cm

Table A2   Reversible and irreversible moisture movement

Material	Reversible mm/m	Irreversible (+) Expansion (–) Shrinkage (mm/m)
Cement-based composites		
Mortar and fine concrete	0.2/0.6	(–)0.4/1.0
Dense aggregate concrete		
Gravel	0.2/0.6	(–)0.3/0.8
Crushed rock	0.3/1.0	(–)0.3/0.8
Limestone	0.2/0.3	(–)0.3/0.4
Aerated concrete	0.2/0.3	(–)0.7/0.9
Glass-reinforced cement	1.5/2.5	0.7
Blockwork and concrete		
Brickwork, dense	0.2/0.4	(–)0.2/0.6
Blockwork, aerated	0.2/0.3	(–)0.5/0.9
Gypsum-based products	approx. 1 mm (based on experience; data not available)	

Brickwork and tiling		
Clay brickwork	0.2 mm(+)	0.2/1.2
External clay tiling	No data quoted. Should be of the same order as higher clay brickwork figure for expansion, i.e. about 1 mm/m	
Sand-lime brickwork	1.0/5.0	(–)1.0–4.0
Wood	Has no irreversible movement	
Softwood	approx. 5–25 mm	
Hardwood	7–32 mm based on 60% and 90% RH	
Plywood	2–3 mm	
Blockboard and laminboard	1–3 mm	
Hardboard	3 mm based on 33% and 90% RH	
Softboard	4 mm	
Chipboard	3–4 mm 65–90% RH	
Woodwool	1.5–3 mm on length	
	2.5–4 mm on width	

Plastics

Plastics are not generally subject to moisture movement but some have progressive shrinkage due to loss of volatiles and related causes; polyesters often absorb moisture and expand, as do epoxides.

Discussion

These tables draw from the Digest the data about a selected range of the most common products used in design. In some cases they were simplified and rounded off; splitting a millimetre is sometimes not significant in building until multiplied several times. Some specific points that came out of the exercise are the following:

- The external temperature ranges which building materials and elements experience are much greater than most people in the industry imagine, 70–85°C for the least sensitive, 85–105°C for metal and plastics, and 115°C for solar control glass.
- The movement of all metals is large, but especially aluminium, copper, stainless steel and lead, those commonly used in building.
- The thermal movement of all rigid plastics is large.
- All cement–bound elements change size continually but slowly as their moisture content changes.
- The figures for clay brickwork assume no restraint of movement but in practice the structure often causes some restraint and this is also true of other materials. The important point is to know what the material wants to do, and to recognize that if it cannot do it, stress will build up in it and may cause damage either to itself or to other things if it is strong enough to affect them.
- Light and dark finishes often make a big difference to the amount of thermal movement.
- Hot weather will quite often cause thermal expansion and drying shrinkage together so that they compensate one another to some extent, but conversely it will also happen that thermal and moisture expansion will occur together and be maximal, or dry/cold shrinkage will happen, causing dimensions to be minimal. *The amounts of*

movement plus or minus should therefore usually be assumed to be roughly additive in moisture-absorbent materials.

- Timber and insubstantial cement-based elements will be particularly sensitive in this respect; for example, the succession of dry and warm seasons occurring from 1974 to 1976 and after 1988 did much damage to things that had previously remained at least marginally sound.
- Plastics and mastics have a gradual shrinkage, accompanied by other degradation, due to loss of volatiles.
- Weather temperatures and moisture conditions at the time of construction have a bearing on what happens later. As far as temperature is concerned, the Digests advise assuming a mid-temperature condition for heat-sensitive elements unless the weather timetable of construction is known in advance. This especially affects the larger or longer elements, metal sheet and trim, pre-castings, long runs of windows and so on. The bigger the total movement, the greater the care needed in design and workmanship.

Deformation

The first Digest of the trilogy is particularly good on deformation, reminding us that temperature or moisture differences within an element will cause bowing, as will combinations of materials with differing moisture or thermal coefficients. It is even likely that differences between white and ordinary cement, or between a surface mix and a body concrete will cause bowing. Twisting and curling are other results of such differences.

Structural engineering

For structural engineering aspects of design the final Digest, No. 229, provides mathematical help with the prediction of distortion.

Reference

1   Allen, W., 'Active buildings – changes of size and shape', *Architects' Journal*, 23 July 1980.

Appendix 2
Background information about polymers and plastics

This appendix provides useful background information on polymers, plastics and related materials without assuming much knowledge of chemistry by the reader and it gives some indications as to how designers should approach the use of polymer products.

It is fairly well known that nature has provided us with a few over 90 types of atom from which molecules can be assembled. Most of the molecules used in the plastics and polymer field are made of only about half a dozen types of atom but they are assembled in many different ways to form different types of molecule. Thus, for example, the molecules upon which polythene (or polyethylene), polypropylene and polystyrene are based contain only carbon and hydrogen atoms, but in different ratios and in different molecular structures. In other polymers, oxygen, nitrogen and chlorine atoms are also used, and just occasionally, fluorine.

Starting with some fairly simple molecules, chemists have invented ways of making them join up into long chains termed polymers. The starting materials are often, but not always, referred to as monomers. Relatively simple molecules are needed to start with (containing between six and, say, fifty atoms) because they can be purified on a large scale by such processes as distillation or crystallization. Impurities even in minute amounts can interfere with the process of making the long chains.

Since the variety of starting molecules is so great, the ways of making them link up become very numerous, and because the modifications that can be made to the finished polymer are also numerous, the possible number of different end-products becomes enormous. All of this is the basis of the big modern industries of plastics, synthetic fibres, synthetic rubbers (or elastomers), paints and adhesives and require a range of expertise now well beyond the ability of any individual to encompass.

With this proliferation of new technologies it is not surprising that some confusion of technical terms has tended to appear and one minor source of confusion in particular is the transatlantic habit of calling some polymers 'resins', short for 'synthetic resins'. However clever the chemists have been in making synthetic polymers, nature actually got in there first. The properties of wood and plant fibres are based upon cellulose, a long-chain material made by linking together a simple sugar-like molecule. Spiders' webs and hair use a nylon-like chemistry. The most 'glamorous' of all is the DNA molecule, the structure of which was elucidated in principle by Watson and Crick in the 1950s. It is not a structural polymer but a carrier of information rather like a magnetic tape.

Fortunately, an understanding of the broad architecture of polymers can be helped by using large-scale models. Once it has been accepted that atoms and molecules can be made to join up in certain ways, the architecture can be represented on the macro scale. This method has been used by chemists for many years and the spectacular demonstration of the chain structure of DNA and the winding of two chains together as a double helix was done by Watson and Crick using quite crude models. It is only when one probes into the problems of why and how atoms and molecules join up in the way that they have been shown to do that one gets into the mysteries and seemingly irrational world of quantum and wave mechanics.

A little desktop simulation can give a feel for some of the terms in polymer chemistry. Paper clips, as mentioned in the main text, can be easily linked together, the single clip being the monomer and the long chain the polymer. As a typical chain can have some 10 000 or more monomer links in it, and a gram of polymer can have thousands of billions of molecular chains, it is not advisable to carry the simulation too far.

The incorporation in the chain of paper clips of a different size or colour gives a simulation of copolymerization, which is a useful way of modifying the properties of a polymer. The use of two different types of paper clip alternately gives an idea of the structure of Nylon 66, but Nylon 6 requires only one monomer unit. Styrene–butadiene rubbery copolymer also tends to have alternating styrene and butadiene units.*

Further experiments can give the idea of chain branching by linking two clips onto the last one instead of only one. Short branches, as well as long ones, are a feature of LDPE (low-density polyethylene) made by the established high-pressure process, and its properties depend on the amount of branching. HDPE, high-density polyethylene, is, by contrast, linear.

The linking of chains together, imagined with safety pins or 'treasury tags', gives a model of 'cross-linking'. Not so easy, of course, at the molecular scale, as one has also to provide suitable reactive sites along the chain and suitable 'cross-linking agents'. In the natural rubber industry the process was originally called 'vulcanization'.

An important factor which is missing with these macro models is the thermal movement which molecules experience at their true scale; Brownian motion of small particles, of which some readers will have knowledge, exemplifies this at an intermediate scale. Depending upon the structure of the polymer and its temperature, molecules are undergoing motion in all sorts of ways – simple translation in space, wriggling, and rotation of various segments with respect to others. The paper clip polymer can be made to undergo most of these movements though with rather limited rotation, while a chain of bulldog clips would be much stiffer at the same temperatures. Indeed it will tend to fall apart if agitated too much, a well-known feature of some polymers. Desktop experiments will also soon show that chains can become tangled together and in fact they are formed in this way in the polymerization process.

At increasing temperatures various modes of molecular motion will be activated and with sufficient jiggling at a high enough temperature the chains can be made to flow past each other under pressure so that useful changes in shape can be made to take place, e.g. as in injection moulding or the extrusion of pipes or sheets.

---

*Butadiene is pronounced buta di-een, reflecting the fact that its molecule has two reactive sites in it.

During such flow, it is easy to see that the long chains will tend to be stretched along the direction of flow. If then the shaped material is cooled too rapidly, there will be many molecules which are frozen in the stretched position. The disorganized, random tangle is the more 'natural' state, so this strain may be relieved later if the fabricated part is warmed, and this is one aspect of the phenomenon of 'plastic memory'.

If the material is substantially oriented, this may give rise to brittleness. In contrast, however, the properties of polyester film ('Melinex' or 'Mylar') depend upon the 'freezing-in' of a considerable amount of such orientation. Rubbery polymers have very flexible chains at room temperatures. If cross-linked to a degree, flow will not occur on stretching but the long-chain segments get oriented. The preference for a random state is the reason why an elastic band reverts to its original state when the stretching force is removed. When an acrylic sheet* such as 'Perspex' or 'Plexiglas' is shaped into an advertising sign or a bath, the deformation is entirely of an elastic nature. Reheating the piece returns it to a flat shape. The sheet is not cross-linked, but the chain lengths are so long that sheer entanglement prevents flow. The corresponding moulding granules are of a much lower chain length and flow in the normal way under heat and pressure.

If the process of cross-linking is carried far enough, the polymer will become a hard, unflowable mass. Heavily vulcanized rubber ('Ebonite') was used as an insulating material in the 1920s and 1930s, and it was used as the front of many of the early radio sets, with colours caused by the excess sulphur used for cross-linking.

The plastics industry has traditionally distinguished between the thermoplastics, which can be recycled by resoftening, and the thermosetting materials of which 'Bakelite' was an early example, where considerable cross-linking occurred during moulding (by compression) and the resulting material was hard and could not be re-used. More modern examples are the epoxy and polyurethane resins.

Another way of modifying the properties of a polymer is to mix it with a plasticizer or extender, usually to make it more flexible. The classic case is PVC (poly vinylchloride) which is a naturally hard, rigid material. To make it flexible, some 50–60 per cent of a small molecule or very short-chain polymer is added. This is cheaper than the alternative of copolymerization with a 'softer' comonomer, which would build the plasticizing effect into the PVC chain. Of course, the plasticizer has to be compatible with the polymer so that it will not separate out, but this is difficult to achieve over the long term required for use in buildings. The plasticizer always has a certain vapour pressure, may be slightly soluble in water (and so vulnerable to repeated leaching), and may migrate under pressure to unpressurized parts of the polymer. Migration from the PVC to adjacent polystyrene can also occur. The author has unhappy memories of a hotel in the North East where the lounge chairs had been upholstered in a highly plasticized vinyl. It was a good trick to get the 'new boy' to sit in one for an hour because getting up usually involved him in getting out of his suit, such was the adhesion between the cloth and the plasticizer that had sweated out onto the surface of the upholstery!

Unplasticized PVC (uPVC) has the advantage of providing rigid components fairly cheaply, but it took many years to develop suitable

---

*A more correct description of acrylic is methacrylic but this bit of sloppiness is tolerated in common parlance.

grades, processing aids, and heat stabilizers to prevent it degrading at processing temperatures.

As with many other plastics, PVC needs additives which prevent or delay attack by UV light and the further degradation that develops from that. The entry of a quantum of UV light energy, which is at the short end of the spectrum, is rather similar on a macro scale to the effect of a direct hit by an Exocet missile on a battleship. It does damage to the chains. Often the easiest course of protection is to load the polymer with carbon black, which has both a screening and a stabilizing effect. If colours or white are required, the problem is more difficult but solutions have been found as for uPVC for external use.

Rubbers, natural and many synthetic types, are especially open to attack by ozone, which is a highly reactive molecule containing three oxygen atoms – in contrast to the oxygen molecule which contains two. As the chemical structures that it attacks are in the long rubber chain, this results in rapid breakdown of chain length and properties. It is especially efficient in breaking the chains if they are stretched (as there is less opportunity for them to join up again) as everyone who has used rubber bands to hold documents together over a long time knows.

The advantage of EPDM (ethylene propylene di-ene monomer) rubber is that the active sites which are necessary for cross-linking are not in the main chain but on a short side-chain so the effect of ozone is not so devastating. There are a number of polymers which seem to have a good intrinsic resistance to UV degradation, methacrylates (like 'Perspex'), 'Hypalon' rubber and some fluorine-containing polymers. Unfortunately they tend to be expensive.

While this 'overview' of polymers may have provided a feel for 'what it's all about', it cannot pretend to help in specific cases. Each corner of the polymer industry has its own extensive knowhow and experience. The range of polymer-based building products on offer is great and manufacturers are often secretive about the exact composition of their products. They have the problem of offering adequate long-term performance at a highly competitive price, and too often considerations of manufacturing cost have outweighed those concerning the long-term retention of properties. With development costs under pressure, evaluation in field or laboratory testing has been known to be skimped and the tendency to regard the building industry as the sink for byproducts, scrap, etc. has sometimes led to the use of raw materials that are difficult to control; hence the emphasis in the main text on the importance of careful manufacture.

Another fundamental problem is that field testing is the ultimate test of endurance, but this takes time and cannot duplicate all the conditions under which a new product may be used. Accelerated testing in the laboratory can go a considerable way to reassuring performance in the longer term, but unfortunately the greater the acceleration of the tests, the less is the resemblance to real life. Especially in the weathering of plastics – and all materials, come to that – one has a complex of chemical and mechanical changes going on at a very slow rate and interacting as they go. At the very least a building requires a 20-year life for products easily renewable and 50 or 60 years for a component that is difficult to replace. The technology for testing and making predictions over such time is still in its infancy.

This 'worst case scenario' may sound rather gloomy, but it should not be allowed to obscure the fact that many successful innovations have been carried out in the provision of building components. If they had not been, we would have been worse off today in the quality and performance of buildings.

So what should an architect or other specifier do when considering the use of a product with which he or she is unfamiliar or which is new on the market? The following is a checklist of possible questions to ask:

1   How critical is the component in the building and what life is required?
2   Has the product an Agrément Certificate? – though it should be remembered that even the Agrément Board cannot cover all possibilities.
3   Has it been used before in the way now proposed?
4   Are there any field test results?
5   What accelerated tests have been applied, e.g. UV exposure, flex or crack propagation tests after exposure?
6   What will be the interaction of this product with contiguous products? Expert advice may be necessary about this.
7   Does the manufacturer have a good track record of innovation?

Arthur R. Burgess, MA, BSc

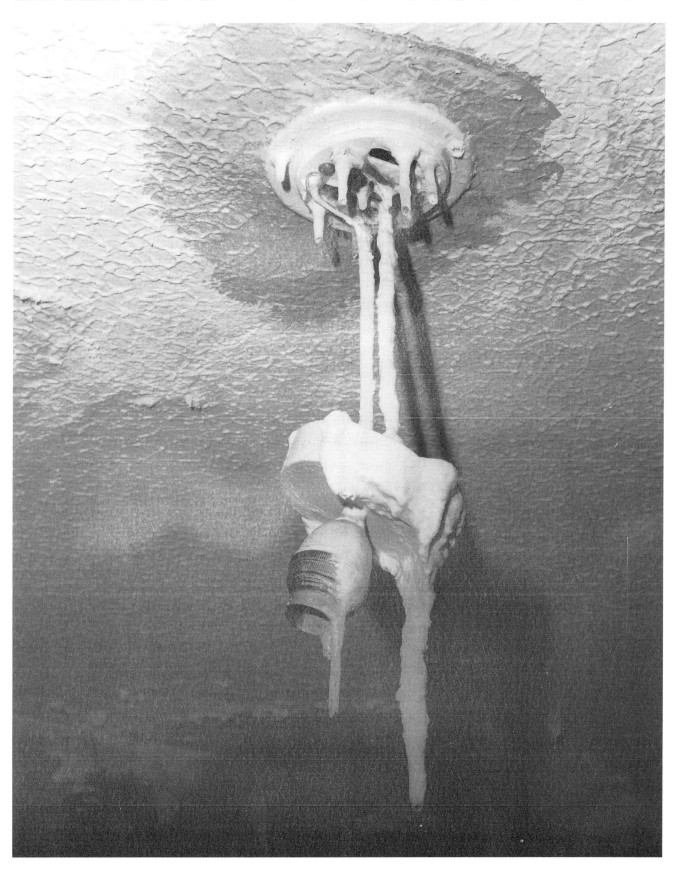

It seemed a pity not to include as a finale this fine photo of that ultimate ignominy, water emerging from the electric system. In fact, the evident alkalinity of the water tells us that it must have been travelling between the concrete and the conduit for at least some of its journey.

Afterword

The vast littoral of industries which make materials and products for buildings is now in continuous motion, most or all of them using science in one way or another to enlarge their particular slices of the huge building market, or to capture new ones created by the depletion of natural resources, or by changing climates, or by engineering developments, or by the onward movement of architecture or simply to correct unforeseen faults. Never again will we know the peace and quiet that prevailed in such matters until well into the 20th century, so what I have attempted to portray here are the physical, chemical and mechanical forces at work on and within the products and subsystems that we use in enclosing buildings so that the evaluation of ongoing change and innovation will perhaps enjoy a strengthened basis. And I have done this largely in narrative form because I believe that by casting it thus as principles, ideas and concepts, and by establishing their relevance to design decision making, a much larger spectrum of useable knowledge can be retained in decision makers' minds than by asking them to carry details of numeracy, physics or chemistry. At its best, that is one of the routes to breadth and authority in imagination.

William Allen

Index

WITHDRAWN